Banking Rebooted

Searching finance

Published under licence 2014 by Searching Finance Ltd.

ISBN-13: 978-1-907720-81-9

ISBN-10: 1-907720-81-2

Typeset and designed by j-views, Kamakura, Japan

Banking Rebooted

John Bertrand

JOHN'S EXPERTISE IN banking, cash management, payments and technology was gained over 30 years at Citibank, IBOS, ALLTEL, Misys and Admertec/Ceptum in the USA and UK.

In this time he created and implemented technology in front, middle and back offices of banks, in addition to developing electronic access options for the needs of corporations and retail customers. This included redefining core banking for Misys (300 banks), ensuring long-term revenue. He has provided and developed cash management consulting, netting schemes, third party cash management, pooling and foreign exchange.

John has been involved in mergers and acquisitions in more than 10 banking/software transactions that included cloud-based solutions. One transaction in the cloud now represents 9% of Misys profits.

John raised capital, added mezzanine financing, leasing and invoice factoring together to create Ceptum Limited, a potential bank, now with $10 million in profitable assets and with no debt.

He has worked with the UK (FSA) and Swedish (FI) banking and finance authorities to create a de nova bank.

Chris Skinner

CHRIS SKINNER IS an independent commentator on the financial markets and, as Balatro, Chris assists clients as an advisor, as well as sometimes being commissioned to write or speak on their behalf at conferences and tradeshows.

As well as being Chief Executive of Balatro, Chris chairs the European networking group the Financial Services Club. The Financial Services Club is a prestigious meeting place, and allows networking between members and regulators, practitioners and futurists. Guest speakers include Charlie McCreevy, John McFall,

Angela Knight and other luminaries shaping the future of banking and insurance.

Chris is well-known for his regular columns and for his daily blog, which can be found at The FinanSer.

He has also written several books on banking, covering retail and investment banking, payments and the supply chain, and general technology trends. Current titles include *The Future of Banking*, *The Future of Finance after SEP* and *The Future of Investing after MiFID*.

About Searching Finance Ltd

SEARCHING FINANCE LTD is a dynamic new voice in financial services, business and economics. Our mission is to provide expert, highly relevant and actionable comment, information and analysis. We bring you the latest industry insight and best practice guidance, provided by writers who are renowned experts in their field, to give you the knowledge that will gain an edge for you and your organisation. To learn more, please visit www.searchingfinance.co.uk

Contents

Foreword

THE INTERNET IS ACCESSIBLE to virtually everyone on the planet, and smartphones and tablet computers are also becoming increasingly commonplace. As people, we now expect 'instant' communication, readily available information on everything, and the ability to pay for something immediately. This is where banks have to be if they are to be a part of the digital age.

The problem is that becoming a Digital Bank is a bit like asking for directions. The awkward answer is that we would not start from here. Unfortunately, 'here' is where we are.

Where is 'here'?

'Here' is where most of the deposit account systems are. Systems that were built before the Internet became part of our lives, but now we have the mobile Internet showing that all of these systems need renovation to cope with the demands of this digital age.

It goes deeper than this – many deposit account systems were built on a raft of ancient codes, lacking a cohesive technical architecture. They are a collection of software packages running on a variety of releases, all with localised codes across the enterprise. Few systems are ever turned off in case something breaks, and it has led to the attitude of 'if it isn't broken, don't fix it'.

'Here' is where we are challenged by change, and specifically the changes required for regulations. Virtually the entire Know Your Customer (KYC), Anti-Money Laundering (AML) and related bank regulations around clients have become law in the last five years. Add to this the new Basel III rules, with its emphasis on liquidity, and the increasing requirements for capital ratios, and intra-day balance movements become a core part of running a bank.

This increases the costs of keeping and moving cash-like securities at light speed, and further highlights the need to digitise the enterprise. Having an industrial age workflow approaching 100% Straight Through Processing (STP) is good, but not good enough to be sustainable in the next few years; it is akin to getting the best out of a steam engine when the world is running on diesel.

The last resistance to digital banking comes from the institutions and corporations they serve. Retail clients are rapidly moving to digital access and the banks that are not servicing them with digital services will lose out. Therefore, institutions and corporations dealing with the digital consumer are becoming influenced by the retail movement.

This book aims to show how adapting to the digital age is not only a priority, but is the only thing that will be sustainable for a bank long term. By moving the data input from the enterprise to the clients on their smartphones and tablets, the workflow is reduced. The overall internal costs drop as activities become more transparent and the likelihood of being non-compliant is reduced. It will then be possible to see what is happening inside the enterprise and focus on how to bring in better working practices and innovation.

Embracing a digital age with a digital bank offer is therefore one of the most critical changes we will see in banking over the next decade, and this book aims to explain that change.

Banking Rebooted

Making money out of real-time banking through the digital transformation

by

John Bertrand and Chris Skinner

Searching finance

Introduction

When the computer or technical device you are working with starts to work in an unfamiliar fashion, the quickest, easiest and often the most effective way is to switch it off and then restart it. That is to reboot the system. *Banking Rebooted* captures the response needed by banks and non-banks to the digital age.

Banks do want and need to be real time, agile and responsive to the new, technology empowered customer. That is us. These customers matter as they have the affluence to purchase a wide range of banking services. The issue for the bank is rebooting the systems.

Marty Feldman, comedian in the Travel Agent Sketch, focuses on a couple wanting to go to the Edinburgh Festival from London but not wanting to travel. Banks with legacy core banking systems, procedures and cultures are in the same mind-set. Yes, they would like to migrate to the new customer-centric, real time banking from the existing 9 to 5 legacy practices but can it wait until next year? This is difficult, as the digital revolution is underway and fast approaching the tipping point.

Revolution is here and when you look back on previous significant changes they start with what we know and then evolve into something more appropriate that is tailored to the underlying change. For example the first motorcar was called the 'horseless carriage'. The first interiors were modeled on that of the stagecoaches. In addition the first horseless carriages had to have a person with a red flag in front of them. So there tend to be two revolutions. The first is the breakthrough, the horseless carriage, and then how it becomes used in our daily lives. Today's cars and their usage are totally different from the early days.

A revolution is underway in how customers spend, move and manage money.

Customers have no interest in how technology devices work internally, as long as they work. Customers have the same view of the internal workings of a bank. All the customer want is for 'it' to work, regardless of what 'it' is. For banks 'it' is seamless, easy and even more self-service banking designed with the customer in mind.

The view should be that of the customer. If the customer's points of contact are handled promptly, efficiently and professionally then regardless of what happens in the back office the bank is rated highly. The key is the customer. So handle the digital banking revolution though the eyes of the customer.

One approach is to provide a customer-centric layer between the mobile/tablet/online customers, usually the banks' most desirable clients, and the past generation of banking systems. So the customer will see a more modern bank, albeit not the finished article, but much closer to where the bank wants to be. As they say, 'perception is often reality'. This layer of capability, a combination of software and people, handles the interfaces between the legacy systems and the forthcoming real time banking systems. It buys time for the bank and retains the customers.

This layer is well known in the capital markets sphere of banking where microsecond responses are the norm and the details of the customers are brought to central location. The first part is for trading. Receiving information on a trade faster than your competitor puts you into a highly desirable position to make money. The centralisation of the customer information came later as the levels of risk and regulator pressures grew.

The layer sitting between the customer and the bank's legacy systems provides the new customer experience, now. The customer sees new banking and new agility. This is good news for the traditional retail and corporate banking where time and practices once stood still.

Gone are the days when you received your bank statement at the end of the month showing you your balance. With digital banking you can see it daily, even hourly if you wish. Domestic payments are down to between immediacy and next day. International payments, bless them, are still the bank's best friend. Money is taken immediately from the bank account and transferred across the banking system. The payment journey often unspecified and frequent running to pre digital timing standards can take five or more business days.

For example, in the 1990s, one of the executives at Citibank in New York bought a yacht from Italy. The yacht was sailed from Italy though the Mediterranean Sea across the Atlantic Ocean and moored in the US. The money was still in transit. No one knew where the money was and this is still happening today.

Core banking engines were designed when banks operated 9 to 5. At the close of day, banking transactions were complied, totalled, put into batches and processed. The batches were sent for the evening's closing of the bank's books, business silo by business silo. The bank accounts updated and the new balances were available, usually around 3am in the morning. Most of the world's banking systems run batch updates to their accounts. In the modern digital world they will need to be online and real time.

Having put the customer interface layer in place, changes have to be made to the supporting systems. Issues such as on-boarding, risk management, transaction processing are also affected by the digital age, and dramatic improvements can be made to these. But these changes should follow after the customer interface has been updated.

This will all cost a lot of money. The *Financial Times* estimates €12.5 billion in new bank equity will be raised in the next 12 months and one of the most pressing needs is modern technology.

Most banks' Core IT systems have been in place for 20 years or more. These systems have been extensively modified rather than renewed over the years to cope with the new and on going changes. But the point has now been reached where this is no longer sustainable and replacement is required.

Changes of this nature are not easy and fraught with risk. A way has to be found to simplify the process and mitigate the risk. One approach to this is to introduce the new customer layer as described, then introduce the new core systems on a selective account basis, in a phased fashion.

Most banks have a high proportion of dormant accounts. Often up to 40% of the client base can be inert. So leave this where it is for a while and move the most active accounts, where the customer and the bank staff will experience the benefit.

Time is the enemy for the existing banking community. It is also the opportunity for anyone interested in banking to become involved in its digitalization. This book is aimed at people working in the financial community and those organisations and people who feel they can do a better job in this new digital world. The interesting thing about digital is it is fast and global and accepted by all.

As the saying goes, fortunes are made at the start and end of empires. Good luck and time to reboot.

Chapter 1: What is digital banking?

FOR HALF A MILLENNIUM, retail banks have worked on the basis of physical distribution. For half a century, that model has been challenged to move towards electronic distribution. At the end of the first decade of the new millennium, we have finally reached the point where electronic distribution has matured, works and is proven. Unfortunately, most banks, and often the senior management and Board of Directors, are stuck in the 20th century. It's time for banks to 'get down with the kids'. It is time to turn the model on its head and focus on electronic platforms, where physical distribution is the cream on the cake, rather than the other way around.

If discussions go: "So, are things like Second Life and Facebook just passing fads, or are they really important to the future of retail banking?" then the questioner is a digital alien.

Digital aliens and digital natives are terms coined by Marc Prensky[1] and refer to different generations of digital usage.

The people who are digital natives are Generation D, the i-Pops ... whatever you call them, they are people who don't think of 'the Internet'. They just get on with their lives and see online, mobile

---

1   'Digital Natives, Digital Immigrants' by Marc Prensky from *On the Horizon*, MCB University Press, Vol. 9 No. 5, October 2001

and all other digital channels as being seamlessly integrated into their world.

These people do not think about branches, call centres, Internet, ebanking, mbanking and so on. They just think of these things as normal, like we think of the car. Imagine life with a car when horses were the thing.

Retail banks have a historically strong branch network.

They added ATMs in the 1970s, call centres in the 1980s, ebanking in the 1990s and are now adding mobile in the 2000s. Each channel is added as an extra layer on the foundation of the branch distribution cake.

Branch networks are the foundations, while electronic distribution is the cream on the cake.

This is why retail banks talk about multichannel strategies where they try to integrate their call centre channel with their Internet channel; they attempt to deliver mobile banking interoperable with the call centre channel; they mess about with CRM to ensure consistency across branch and Internet channels.

The problem is this: banks only have one channel.

They do not have multichannels, call centre channels, Internet channels, mobile channels and so forth. They just have an electronic channel that underscores and provides the foundation for all end points: mobile, telephone, Internet and branch.

This is the big change: banks should stop thinking of channels and just recognise that they are digital enabled.

Call centres, ATMs, branches, Internet, mobile ... everything is digital enabled and therefore the bank has become a Digital Bank, based upon digitised platforms that reach into every nook, cranny, sinew and synapse of the bank. This has happened as a reaction to market forces over time and most of the banks still have everything built in layers of complexity and legacy.

The ATM, call centre and Internet channels were all built as layers of cake created when the physical branch was the foundation. The electronic channels were built as ancillary to the core branch channel. That is why they were often separated and have this chasm of non-integration between each other, as banks were built on a physical distribution model where electronics were layered on top. To facilitate the many ancillary systems alongside the core system, necessary middleware has been introduced. Yes, you have guessed it – now many banks have many ancillary systems and many middleware solutions. Not ideal even for end of day processing, let alone intraday banking.

However, today, and certainly tomorrow, the population has moved to a world where the majority are digital natives. As the digital generation age matures, what role will there be for banks that have been built upon the basis of a physical distribution model with electronics layered on top?

It's time to turn this on its head. It's time to think about banking as an electronic structure. It's time to bite the bullet and admit that retail and commercial banking is not a physical distribution structure with electronic channels on top, but an electronic distribution structure with electronic and physical channels on top. It's time to become a Digital Bank.

This means wiping the slate clean and starting afresh.

How would we build today's bank if digital networking is its foundation; and call centre, Internet, mobile and the branch are just the cream on top of the cake?

Where would you build branches, and how would you build them, if the branches are ancillary and perfunctory to the electronic foundations?

Who would you employ, and how would you employ them, if the core differentiation of the bank is its digital base rather than its branch structure?

The fact is that any bank that launched today as a greenfield operation would think this way and, with the right leadership and implementation, it would thrash the weak competition based upon legacy structures and legacy thinking that exists in most markets.

Start thinking about the bank. The bank has to be digital at the

What is digital banking?

core with layers of distribution on top and branches as the cream on the cake.

It's time for change.

Designing the Digital Bank

AS BANKS DESIGN THEIR NEW GENERATION Digital Bank, the start point has to be customers, employees, processes and organisation structure; then secondly, the need to consider how to build the optimal way to use digital resources to reach them; and finally, how to build or use traditional bricks and mortar to manage the organisation structure at the end.

Banks are trying to do the latter and, with greenfield operations, could do so brilliantly. Instead, due to the fact that they started building with physical structures years ago, banks have to find a path to marry the two worlds. They are achieving this, in part, by building their digital architectures around the rewiring of the existing buildings that they want to keep in play.

The most important consideration here is the digital architecture. What does this mean in practice?

It means that banks need to recognise that they have been built in stages in often siloed states. Often with core systems that are pre-Internet. This current state needs to be reconstituted and reconstructed for digital processing. A template, if you like, that is put over the bank to show which parts need to be reconstructed, which remain the same or which are shut down. Banks rarely turn off systems; instead they let them tick away, paying the maintenance, just in case they have been wired in to a working system somewhere else.

As a digital business, all banking can be broken down into pure bits and bytes but, more than that, a bank can be seen as three digital businesses in one.

It is a *manufacturer* of products; a *processor* of transactions; and a *retailer* of services.

In this context, the digitisation of banking becomes more interesting at a strategic level.

Deconstruction

EVERY BANK PRODUCT can be deconstituted into its lowest common denominator of components, and then reconstituted into new forms of use and structure. In other words, there are no integrated product sets any more, but just banking as apps that customers put together to suit their own needs.

Bank products are becoming apps, manufactured in such a way that customers can put them together to suit their lifestyle and their businesses.

The open sourcing of digital processes is rife and has disrupted and changed everything, from how operating systems operate vis-à-vis Linux, to how Google develops its omnipotent reach.

Learning from such open source processing, PayPal launched X, a developer-based service for PayPal processes as APIs – Application Program Interfaces. APIs are the interfaces into systems and each one is tailored to a particular function. The good news is banks have APIs and are willing to share them.

PayPal's API allows anyone to pick up and drop PayPal into their systems and, like banking products as apps, allows PayPal to be reintegrated by third parties into any code and operation desired.

The result is that PayPal's relevance increased massively overnight and led to Citi following a similar approach, when they announced that their transaction services would be offered as APIs at SIBOS this year (SIBOS is an annual conference, exhibition and networking event organised by SWIFT for the financial industry).

Bank processing is open sourced coding, offered to anyone to plug and play with their offerings through APIs.

Finally, the customer relationship has also changed.

The customer relationships used to be human, one-to-one. Then it became remote, one-to-many. Now it is digitised, one-to-one, with the customer being able to self-service their banking needs once they have identified what they require. The bank still has to give approval, but the traditional administration has been automated and moved out of the bank to the customer. Like pumping gas, we now all do it ourselves.

This is where Big Data comes into its own, as we are now trying to manage remote relationships leveraged through mass personalisation. Mass personalisation can only be achieved by offering contextual servicing to each and every customer at their point of relevance.

This means analysing petabytes of customer data to identify, on a privacy and permissions basis, what contextual service the customer may need as they live their lives.

If they are walking past a car showroom, do you promote cheap motor insurance, or a car purchase scheme?

If they are leaving the casino, do you offer a loan, or a referral to an addiction clinic?

Some of these may seem controversial, but we are already seeing contextual offers through finance coming into play in the form of Google Wallet. For example, as Google tracks your searches for plasma TVs, you get an offer for £200 off the TV you spent the longest time studying online as you walk past the electronics showroom today. But the offer is only good for an hour, and only as you are in proximity of that electronics showroom.

This is the new augmented reality of customer intimacy through Big Data analysis, and bank retailing and commercial businesses will be based upon the competitive differentiation of analysing mass data to deliver mass personalisation.

In summary, the digitisation of banking will be packaged as digital structures where products will be apps, processes will be APIs and retailing and businesses will be contextual, monitored and delivered at the customer's point of relevance.

What does this mean for the physical bank?

BECOMING A DIGITAL BANK, with digital networking at the core, is a real challenge as it means moving fundamentally away from placing branch networks at the core.

Some people believe this is purely academic as we have branches today and can't get rid of them, so the question is how to use these branches.

It is obvious that branches are critical sales centres and, in the future, they will not be transaction centres. Australian banks now charge $5 per banking transaction that could have been done by the customer. The customers immediately got the message and mobile phone usage has soared. Historically, branches have been transaction centres. This is what everyone is struggling with today: the move from impersonal processing to warm client relationships.

Transformation

THE TRANSFORMATION PROCESS STARTS with moving branches away from transactions. Transactions can be managed through remote telephone and Internet connections and self-service machines. The human contact should be as local as possible, whereas texts and emails can be performed anywhere internationally. Once the transactions have disappeared from the branches they are in locations which are convenient to the customer. Branches are often in places where they are not convenient, especially in the digital age.

This is why some bank strategies are fundamentally flawed, as those who think branches are the start point will throw good money after bad. Those who think digital networks are the start point, and then build the end-points on top, which includes bank branches, will be much nearer the right strategy for the future.

A bank strategy today needs to start around a digitally enabled bank. If you were designing that bank, then here's the question to ask around the branch focus:How many branches would you layer on top?

❏ How many of those would be self-service automated branches and how many would be sales centres?

❏ How many and what sort of staff would be needed to relate to the new era customers?

❏ What are the technological aspects of the digital enabled branch in this context, and how much technology do you put into the branch?

❏ What is that technology doing and how does it profile against the staff skills and customer demographics?

❏ How does the underpinning of the new digital enabled branch fit with the digital enabled alternative contact points?

It has to start with the network being the IP (Internet Protocol) network of the 21st century and not the high street bricks network of the 20th century.

Bank design versus architecture

THE REASON WHY THE FOCUS upon becoming a Digital Bank is such a difficult one is that people often confuse design with emotion, architecture with distribution, and channels with infrastructure.

The design of a Digital Bank begins with architecture. The discussion gets confused with bank design, which is different. Architecture is about materials, dimensions, frictions, structures; design is about the user experience, the customer engagement, and the human connectivity, whether it is by face-to-face or screen-to-screen.

The two go hand-in-hand. The bank designer would start with the customer and how to focus upon customer emotion and behaviours. We call this 'buyology'.

Although biology is a well-known science, there is a lesser known but increasingly important science that is also, by coincidence, called 'buyology'.

Buyology is a core science for designing banking in the new world of freakonomics, where everyone is struggling to understand the methods to get customers buying, and is defined as: 'the science of understanding business relationships'.

Buyology is all about knowing why people buy. How to create business encounters where purchases are made, that can be replicated over and over. It is the ability to create long-term business relationships, not just a one-sale stand.

Bankers are learning buyological processes because traditional selling and advertising no longer work. People don't want to be sold to and they certainly don't believe corporate speak. What customers really want is to deal with banks that demonstrate a true understanding of their individualistic retail and business wants and needs.

Banks' understanding of buyology is a clear strategic imperative because business has become so transparent thanks to blogs, Facebook, Twitter and other social media. These networks now ensure that any cover-up of any issue will be exposed. That is 21st century Internet-enabled consumer power. Social networks mean that banks must start demonstrating clear integrity to be trusted and be fully transparent.

Bank relationships are based upon trust and trust is easily broken. The relationship has become one where the trust is in ignorance. A little like a father-child or priest-confessor relationship, the customer has to believe the bank is looking after their best interests. Unfortunately, this is being called into question, thanks to the new regulations implying that banks don't always act in their client's best interests (really?!).

This trust has also been tested by Enron, Worldcom, Parmalat and such like, and is being tested again in the credit crunch. In fact, the recent admission by the Bank of England that it didn't understand the financial markets anymore, in light of the Northern Rock collapse, is shocking. When the regulators and co-ordinators of the financial markets lose their understanding, something has to change.

What is digital banking?

In the digital age the future buyer will not buy from those they do not trust or understand. They will instead use the power of social networks to find the truth and will move towards those who deal with integrity. In other words, buyology means knowing your customer so well that they are no longer a customer, they're a partner.

Relationships are based upon understanding and compromise. We talk CRM, but you don't have relationships with customers. Customers are sold to; partners have relationships.

Banks that turn tellers into sellers, or have massive egos in the dealing room, will soon find that the truth will out. Instead of sustainable sales, they'll get lots of one-sale stands. The real partnerships based upon fair dealings with trust will be the long-term sustainable relationship businesses.

If you accept that the future of banking will be based upon which banks are the best buyological scientists, then this becomes the premise that the bank designer would use to build the Digital Bank.

The bank would be based upon digitised techniques of customer understanding to build processes from the customer viewpoint. At the end of designing, the bank would then go to a digital architect to build the digital design.

The digital architect

BUSINESS PROCESS RE-ENGINEERING (BPR) is about process re-design and not to be confused with design itself. BPR makes the running of the steam engine more efficient; it does not change the fact it is a steam engine. The digital architect is needed to design the bank for the digital age.

Architects have been called in recently because banks' foundations are suffering from subsidence. Foundations were built on cement and bricks-and-mortar, and those foundations are cracked due to the revolution in technology over the last five years. Most banks got away with painting over the cracks but, today, they are finally saying they want to replace the brick foundations with technological foundations in the form of digitised architectures.

The architect is there to replace the physical foundations – process implementation – and the designer is there to work out what the new bank house should look like – process redesign.

Likening this to the building trade illustrates the point well. A house or building has foundations. Bank architects of the last few decades used branches as those foundations, but today instead would use IP infrastructures.

This does not mean that branches or people are obsolete. The branch and face-to-face discussion is more to do with what type of house you want to build. In other words, it's the design, the vision, the interior decoration, the furniture and the other bits.

The designers may say: "I want to build a high net worth house, with sales advisory centres for people who want face-to-face engagements". In this case, you build your bank house with IP foundations and lots of snazzy advisory centres, or branches, in the physical world. Others may say: "I want to build a low-cost high volume processing house, with minimal physical contact", in which case you build your bank house with IP foundations and hardly any branches in the physical world.

Either way, the IP is the foundation, and this is where the architect will start.

An architect of today's bank starts with technology as the raw material for the building, especially IP networking.

The bank must then work out what customers are sought and what those customers want. Build the bank and design it based upon desired customer experiences. Build those customer experiences to appeal to the behaviours of the target audience. Address the needs of digital natives, digital immigrants and digital aliens, and work out how the designs address this mixture of customer types. What experiences and behaviours will these different audiences require and how is it best to deliver them?

These are all designer questions and nothing to do with

architecture. Once you have your design you can then give this to the architect to work out how to build it, and the architect will begin with a base design using IP as the foundation (business and IT working together).

The reason people get confused about this segregation, the channel mix, the house design etc., versus the foundations of the bank, is because they mix up process redesign and process implementation. The channel strategy is the house design; the building strategy is the architect's digital materials.

The focus must move to a strategy where the architect lays digital foundations, rather than trying to maintain the old brick foundations. It's to do with the materials at the base of the bank and the fact that these materials are fundamentally different today, because they are digital rather than brick-based. This is why banks need to fundamentally redesign.

This redesign is to replace the building foundation. In replacing the foundation, the strategy for the design of the house itself may also change, but this is still very much open to the designer's competitive strategy. It is a totally separate discussion that has nothing to do with architecture.

The architect is purely working out how to replace the foundations with IP. Therefore, the designer's role is to tell the architect what the designer wants to build on top of the foundations – the business the bank wants to be doing.

The two roles – architect and designer – go together but are very separate and distinct roles. The reason the redesign started in the first place, however, is because the foundations are crumbling (the branch brick-and-mortar model) and needs replacing through a new architecture – IP networking.

In conclusion, most banks have their foundations in branches as the raw material, and that is forcing them into poor designs that do not match the way they want to behave. This is why they are hiring architects to replace those foundations with IP. The architects are then asking the bankers: "What design would you like to have on top of these foundations?"

The Digital Bank demands a Digital Bank

THROUGHOUT THIS CHAPTER, we have focused upon designing the Digital Bank, evolving from the physical bank, and recognising that the new bank is very different. It is deconstituted and needs to be reconstituted; it is modular and plug-and-play and no longer integrated or end-to-end; it is remote and human and has digital at the core and the logic flows from that core.

This then leads us to a very different but clear view for the future Digital Bank. The theme is how you turn a vertically integrated business that owns the customer process end-to-end and organises itself around products and channels, into a horizontally-structured business. That business wants to provide functionality to the customer at their point of need and organised around the customer's data.

Silos-r-us

BANKS WERE FIRST CREATED to look after all the financial needs of people and businesses. They were licensed to live in their own segregated world of operation and completely owned that piece of turf. Everything from taking deposits to giving loans was the banks' domain and they were organised to do just that. Often government protected and with stringent rules that limit competition, regulations are such that they are becoming a bank's USP.

As a result, most banks created operations based around products: money transmissions, mortgages, cards, loans, insurances, etc. These were delivered through one channel – the branch.

Over time, another channel appeared – the direct sales

representative. These sales folk resided in branches and were served by the branch system. Then, a new channel popped up – the call centre.

The call centre was like one massive remote branch and required a new structure to operate from. But the underlying data could be delivered through the branch-based systems, so the new structure was primarily designed to sit on top of those systems, offering scripts into the various products the bank offered. The call centre people struggled with this – sometimes operating six or more windows of screens at any one time to get a competitive picture of the customer's needs – but they lived with it.

Then, another channel popped up – the Internet. At first, banks thought this could lead to branch closures and started to invest heavily in moving from branch to Internet services. However, the underlying data was still held in product silos and the Internet was not responsive to customer's views of the world. Broadband had yet to appear and customers were reluctant to lose their branch connection.

So, the banks left the Internet as another layer on top of the branch-based systems, alongside the call centre spaghetti. Banks had become locked into vertically-integrated processes, structured around product silos that were ill-suited to the multichannel world they now served. But it was ok. Using middleware, fudge, smoke and mirrors, it did the job.

Then a perfect storm of mobile, cloud and big data appeared, augmented by customers tweeting and socialising 24/7.

Now here's the challenge. The bank cannot leverage data; it's locked in product silos. It cannot serve the customer's needs. Banks layered channels over products. Now, they need to leverage data over mobile. And banks lost the end-to-end process as customers moved to apps and pieces of process and functionality as needed. Now there's a need to organise the bank around the customer's data and then leverage that data through the cloud to mobile devices as apps.

Time to de-silo

THE WAY FORWARD is to completely rip out the old systems and replace them with new core banking that can service the bank, and therefore the customers, in the way that is appropriate for the 21st century. Well, banks are starting to do just that. Some are having problems, but this is why banks are changing core systems. You cannot restructure a bank around customer if you have that data locked into legacy systems that are product-siloed and channel-handcuffed.

Chapter 2: The implications for retail banks

DURING THE PAST FEW YEARS, the world has moved into something loosely described as Web 2.0.

Web 2.0 is applied to the age of easy generation of content, media, activity, networking and more, and these combined developments of technology are pushing banks in the 21st century to create the Digital Bank.

The technologies themselves have been emerging over decades, but they are especially hot now because they are reaching maturity. The technologies fall into four major categories:

❏ Mobile networking and social technologies;

❏ Data analytics;

❏ Unlimited networking and storage;

❏ Modular computing.

Mobile networking is all about the emergence of the mobile Internet, and the combination of telecommunications with information technology.

Social technologies build on this, but are focused more upon the developments of consuming content on the Internet in the 1990s to creating content in the 2000s onwards. Fuelled by Twitter and Facebook, the social Internet is not just about these platforms but about the plethora of methods of creating content from consumers to corporates in a simple, interactive, social manner using 21st century technologies.

Data analytics is briefly labelled 'Big Data' today, but evolved from the data analytics and data mining techniques of decades before.

Unlimited networking and storage applies the idea of Moore's Law (double the power at half the price every 18 months) to all areas of computing and has led us to the cloud-based structures of today, where anything can be stored in any file size anywhere on the Internet, with easy access in nanoseconds.

Finally, modular computing points to the developments of object orientation and plug-and-play computing to enable banks to become component-based APIs – Application Program Interfaces, a form of packaged functionality – on the mobile Internet.

Here are a few illustrations as to why they are such transformational technologies.

Mobile is truly transformational as it has moved us from having to go somewhere to do something – a physical place or a desktop screen – to having connectivity in our pockets and purses. This means the 24/7 availability of service is now the required state rather than the dream.

However, it goes beyond this, as mobile delivers two further transformational moments.

First, it gives everyone on the planet the ability to communicate. In the farthest reaches of the world, people are communicating wirelessly in a way they never could before.

And you may think that these remote corners of the world just have simple dumb 2G phones, but you would be wrong. The fact that most affluent consumers change their mobiles every 18 months means that emerging markets are getting smartphones sooner than you think. Affluent consumers usually mean successful businesses and that means money, and banks look after money.

In other words, everyone is socialising via mobile. Anyone anywhere can now relate socially and communicate globally. Seven billion people globally are connecting one-to-one, person-to-person, peer-to-peer.

Second, mobile provides a transactional infrastructure that was non-existent just a few years ago. It is the reason why Africa has seen the most rapid transformation through mobile, with M-PESA in Kenya regularly cited as the most revolutionary change.

This means that communities that once only had physical connectivity now have digital connectivity. Communities have exploded from local to global, and the wireless transmittance of anything to anyone, anywhere is a reality.

Mobile is part of the reason why *Social* is now a major shift.

After all, communicating with friends and family is what most people use a mobile telephone for, and the ability to create and share photographs, updates, news, links and more via mobile is the truly social hemisphere of tech change.

Add on to this that mobile allows you to locate anyone, anywhere, any time, provides a further dimension of change, and brings in the importance of *Big Data* and *Cloud*.

Geolocating targeted consumers with offers at their point of consumption is a big piece of change. This is the dream of marketers, and is now a reality. Forget mailing coupons in the post, you just say at the point of retailing: "Here's the deal".

But you cannot do that without having massive analytical capability of data to dig out what is relevant to whom at what point of time.

Big Data, storage and mobile brings all of this together with the Apps.

So these four technology shifts are centrifugal forces of change today, as they are massively complementary.

Mobile allows anyone to *socialise* with anyone on the planet, whilst *storage* allows companies and government agencies to sift through the massive amounts of *data* that the *mobile, social* world is creating.

Chapter 3: The implications for corporate banking

THE KEY FOR CASH MANAGEMENT in this era is to make certain accounts do not go into negative territory, as most banks charge fees for credit provision, often up to 20% per annum, but the real killer is the 'unauthorised overdraft' charge. Here even a slight overdraft can incur punitive charges. For example, the UK banks charge fees that can easily exceed 'pay day loan' charges.

Banks have also taken a zero tolerance approach to credit, with many reducing the credit line available and/or increasing the interest rate being levied. Credit from banks is scarce and has opened the doors for non-banks to fill this role.

The first thing is to make certain the bank accounts, wherever they are, report in first thing in the morning to a regional or central location and the time for a Digital Cash Map is here.

Why?

Because the key attributes that make the treasury function perform in the top quartile for cash management are automated reconciliation of the bank accounts and Straight Through Processing (STP).

Equally, cash and liquid instruments remain the most critical tools, but the challenges are: where is the cash in the enterprise; how up to date are the cash flows; who has paid or not been paid;

how to bring the cash together for the best liquidity and investment opportunities; and so on.

The cash management process

BACK TO OUR DIGITAL CASH MAP and we need to go back to the origins of cash management, which starts very simply with a bank account. The company opens the account with a bank to begin business. For any company, the initial reason is to pay suppliers (creditors) and receive funds from sales (debtors). These two activities are termed accounts payable and accounts receivable, and are the financial centre of the company.

Cash management starts simply enough, but soon grows in complexity as the company grows and changes. The first bank account is usually registered in a single currency. As the company grows and expands, the company will often rapidly transact across borders to meet the increasing demands of a global audience through the Internet, and now the mobile Internet.

As soon as a company opens a bank account with a different currency they have moved into the world of international cash management. Before that many companies bill in one currency, usually the US dollar. This places the foreign exchange conversion costs onto the buyer and simply adds an additional hurdle in making the sale. Going digital means a bank or non-bank can provide the currency exchange needed for the buyer so that they can use their own currency of choice.

By taking a second currency, the corporate has to consider ongoing foreign exchange, exposure management, netting, pooling, tax and regulatory environments. The level of cash management complexity has exploded and continues to accelerate as the company adds more international business.

What this now means is that companies have to rethink their cash management as it is now multicurrency. What to do?

The first step is local cash management often referred to as 'domestic cash management'. Once another currency becomes active then the step becomes international cash management, followed by the final step of global cash management.

It is the addition of the second currency that not only layers in a new set of services, but also a whole new set of fees as banks have historically earned considerably more from 'international' transactions.

International payments, for example, are often ten times (10x) more expensive than domestic ones, and these payments have to pass through another country's payment infrastructure. This infrastructure often requires additional information and changes to the format of the messages. In some cases, the domestic payment infrastructure is so old that truncation of information takes place and shorter information than the overseas payment systems require, is sent or received. This is why international payments can take far longer and also why these payments increase the float to the banking system.

The European Union is creating the Single Euro Payments Area (SEPA). SEPA's aim is to standardise payments across the euro zone, and mandates to the banks a set of standard procedures to enable the euro to act like a domestic currency across EU country borders.

As each country has its own existing and working payment settlement systems operating under disparate regulations, the change needed to meet a global standard is massive but is slow in coming. The UK, one of the most advanced countries in real-time payments, has yet to move to the standard ISO 20022 message types. The UK payment schemes move at the pace of the slowest bank (the government requested Faster Payments in 2005) and finally in 2009, with EU law on euro payment time and mobile payments delayed until spring 2014. Well, the UK banks do own the payment infrastructure!

One aspect of the SEPA is the adoption of standards on a global basis, which can be seen by the increasing use of the IBAN (international bank account number) and BIC (bank identifier code).

The implications for corporate banking

Each bank has established its own BIC, which is SWIFT-compliant (SWIFT is a a banking community-owned infrastructure). With the IBAN each company's bank account can be readily pointed to and payments directed from any account, anywhere. Only the ability to digitalise these long streams of numbers will enable cross-border payments to be made from payee to payer across the bank payment landscape.

But cash is not the only challenge as most companies have more than one banking relationship. Multibanking occurs as the corporation grows because a growing company attracts different banks in different countries. Both the bank and the corporation do not want to be reliant solely on each other. Consequently multinational and very large national companies have more than one bank.

The first digital cash management ethos is to make certain the probability of any bank account going into an overdraft position a virtual zero.

Banks are starting to make digital banking the preferred approach anyway, as information comes and goes in a fully automated format. This makes it easier to receive and deliver information to the client. In addition, smartphones and tablets are beginning to be the devices of choice for many. The trend amongst a number of banks is to merge Internet banking and mobile banking into digital banking. Many of the Internet banking services were created for the corporate market, when banks provided the entire infrastructure, and are now being decommissioned. There are also other significant impacts. For example, by 2014 there will be more Internet-connected devices than people. All of this means that there is a growing need for real-time bank-to-corporate connectivity that is industrial strength and available to all.

Historically, the banks that have invested in Internet banking have done so in such a way that the interface between them and the customer is unique to that bank. Consequently, using more than one bank makes consolidation at the company another necessity, using two or more interfaces because none of this has been standardised. In fact, few countries have insisted on a multibank format or interface standard, with the notable exception being Germany. German banks operate in a multibank environment, so only one interface is needed for their banking community. The need to build interfaces restricts the flow of information within a country, forcing customers to only use one bank purely for ease of use.

The move to a fully digitised state is assisted by using a bank's API. These APIs are specifications on how software can interface with the existing IT structure, and allows corporate treasury activities to be in synch with the bank.

As the company grows, the need for additional bank accounts becomes a necessity. The accounts are opened for a variety of reasons, but the need to make the most out of the money is still there. It is here that the complexity of cash management begins to increase and banks begin to specialise.

The first step is to bring the accounts of the company together. The oldest, simplest form is the mailing of the bank accounts to the same address, or showing the accounts as separate accounts on Internet banking systems. The bank's systems often do not know if the accounts belong to the same company. If the postman or the mail man is delivering bank statements to your company the chances are the company and/or the bank has yet to embrace digitisation.

Paper is the enemy of efficiency in the digital age

BANKS WANT TO REMOVE INEFFICIENCIES, as it costs banks to repair, for example, international payments if they go wrong. In fact, reconciliations and exceptions management for international payments is a small industry in itself, with the banks incurring the expense.

It is here that the technology and processes applied by certain banks start to earn them a name for cash management. Both the technology and processes are moving towards industrial strength with non-stop availability. In addition the availability of inexpensive software – apps – to assist consumers, and soon companies, in reconciling their bank accounts has made the bank client keener to have more sophisticated techniques – techniques designed to save money and reduce errors.

The starting point for these efficiencies for the corporate is to add BIC and IBAN numbers to their client invoices, as more information is needed on invoicing to international clients and suppliers. As BIC and IBAN are long streams of numbers, the chances of them being accurately transcribed manually are low. Hence the corporates and the banks are changing to provide this information electronically.

Equally, on the international side, SWIFT provides data from other banks using a dedicated message type. Companies can receive their balances and transactions from multiple banks as long as their bank offers this service. In turn, the companies' international banks should also have installed the SWIFT functionality.

This is leading to competition from banks outside the domestic country for the larger companies' banking business.

For example, one key technique is the automated movement of funds between all of the accounts, so that no single account generates overdraft charges. When one of the accounts is overdrawn, another account has more than enough to cover the shortage and the systems move the funds accordingly.

It is similar on the investment side, where consolidating money enables the company to generate a greater level of interest payment and return to the company as banks pay more for bigger amounts. In addition, the use of digital nominee accounting, where monies can be offset across geographies with no money being moved, is becoming mainstream. Finally, the level of transactions by accounts are totalled and matched against an agreed level of transaction charges. It is the companies with strong transactional business that are valuable to banks. Providing the credit situation between these corporations and their bank is not locked through contract or proprietary technologies, they can offer their banking business out to competitive bids.

The final development in this area of automating cash management is the creation of the best blend of domestic and international cash management into global cash management. Global cash management is the province of the top 10,000 corporations. It is here that the most sophisticated of cash management services are developed and applied.

With digitisation, this sophistication is not just the preserve of top companies. An early example of this move from only the elite to common place is the sweep account. Designed initially for the larger corporates to move money automatically into a money earning account, it has now become widespread across the banking spectrum.

There are also alternatives to using bank accounts for corporates, such as using warehouse accounts to enable money movement anywhere.

Warehouse accounts are held by non-banks, and offer an alternative payment process to the banking structure. The largest provider of this approach is PayPal. With hundreds of million accounts worldwide, PayPal competes with the traditional banking payment channel known as 'correspondent banking', to offer money movements for any company or individual.

By using a warehouse account held by PayPal and similar companies, money is lodged in that account by transfer from a traditional bank account or credit card. The holder of that account can then move money to another person with a warehouse account with the benefit of prompt and convenient payment.

Whether a warehouse account or a third party payments provider or a bank, removing paper and inconsistent standards from the international banking system removes tremendous inefficiency, which is good for commerce.

It is not necessarily all good for banks, however, as removing the inefficiencies of the payment processes costs the banks money in reduced float, where money is made by having money on account.

Over time, the pattern of money flowing into the bank account becomes predictable. Often a company has peaks and ebbs in its accounts. The account can be reviewed to see which moneys can be invested for a longer period. Banks often pay more interest on money put on deposit for longer periods. Hence corporates with established balances can move money from day-to-day savings into longer term – 7 or 30 days or 3 or 6 months.

The area of forecasting money both into and out of the account has to date received little attention. Most banks do not offer any services around this area of cash management. The corporates are also guilty of not attempting to forecast cash flow for accuracy.

The keys for forecasting cash flows are revenue and expenses. The expenses side tends to be the easier of the two to predict. The greatest cost for most companies is their people. Given many countries' employment laws, these costs, depending on the length of service the people have with the firm, should be considered with a minimum of three months' break. In most cases when layoffs are required, there is money set aside to meet the financial redundancy needs. So simply forecast the payroll as it is today for the next 12 months, adjusting each quarter to the changes in the company's head count.

Revenue is the most unpredictable as is the cash arriving into the bank account. It is common to see company financials on a quarterly basis compared with the same time period 12 months previously. As the money is the procession through the financial year, comparisons can be made between similar periods: half year compared to half year 12 months ago is typical. Similarly, cash flows should be forecasted. This way, the next quarter's cash and 12 months rolling forecast can be reviewed and actions taken to improve cash collection and make determinations on the cash distribution. In addition, the business can see potential shortages or excesses coming.

The revenue component can be split into two – recurring revenue and all other. Recurring revenue is common in many sectors, for example insurance and software. Here clients have agreed to pay monthly or quarterly revenues tied to signed contracts. (The best contracts are those that exceed three years as they can be securitised and used as collateral should additional funds be required). Most companies have a combination of the two. The next step is to turn this into predictable cash. The best way to do this is to sign up a direct debit. Not only does this make the cash 95% predictable, it also guarantees the cash will arrive on a certain date.

Chapter 4: The increasing use of mobile

WE ARE SEEING MANY CHANGES in the world of banking, but none as revolutionary as in the mobile space. There are more telephone handsets in use on the planet than toothbrushes, and developing economies such as India have more mobile telephone handsets than toilets.

The world has a handset in existence for every person on the globe. Although that doesn't mean everyone has a telephone, it does mean that everyone has the potential to access a handset.

This is revolutionary, baby

IT MEANS THAT THE WHOLE PLANET is wirelessly connected. Almost every person in the world has the potential to electronically interact with someone else. That has never been a capability in our history of existence, and is the reason why this is such an exciting change.

For banks, this is revolutionary as these changes are allowing many new and innovative services to be launched, from direct Person-to-Person (P2P) payments via text messages, to contactless apps and dongles for payments, to full-scale financial services delivered over tablet and smartphone systems. It is the reason why PayPal and Google have invested over $1 billion in mobile payments development between 2008 and 2012.

According to Gartner Group, the mobile payments market is hot because it produced around $235.4 billion in global transaction values in 2013, up 44% from 2012. That includes money transfers, bill payments and merchandise purchases.

This has created some challenges. For example, cut-throat competition and poor risk management, as some of the newcomers are not screening their business customers effectively, raising the risk of mounting losses.

Between 2008 and 2013, venture capitalists invested over $1.5 billion into the mobile payments market but have little to show for it, with only Square being a major success.

Square was founded in February 2009 by Twitter co-founder Jack Dorsey. The system is a simple idea to put a payments dongle into the earpiece of a mobile smartphone. Once inserted it allows the phone to become a point-of-sale and take payments. Ideal for small merchants, the system slowly grew and then accelerated rapidly in 2011. For example, it took about 10 months from its launch for Square to reach $1 million a day in payments transactions going through its mobile app. This doubled to $2 million a day two months later and one month later, in May 2011, it was generating over $3 million a day, resulting in Visa making a major strategic investment in the business. Visa can export this technology across its network and pull in more fees – usually 1.5% of the transactional amount.

Banks have been active in the mobile payments space for almost two decades. Back in the 1990s, banks were playing around with Wireless Application Protocol (WAP) for account balance checking on mobiles. However, much of this was experimentation and it was not until the late 2000s that mobile became really hot.

Simple mobile systems

As mentioned, we have moved from being a disconnected planet, where few had access to technological links, to a world of connections where everyone can wirelessly communicate. For Africa, Asia, Latin America, China, India and other developing markets, this is where we can see the true revolution in both society and banking.

Cheap access to mobile communications has given Africa and other economies the ability to transform from a world where crime was simple and payments were hard, to one where crime is tracked and payments are easy.

This is because electronic payment processing is both traceable and immediate, with no challenge of distance, and the best example of such change is seen in M-PESA in Kenya (although this is not a typical implementation of mobile payments).

M-PESA – 'M' is for mobile and 'PESA' means money in Swahili – was launched in September 2007 when the Kenyan government asked Safaricom, a division of Vodafone, to help improve the way money was moved between citizens. At that time, most payments from towns to villages were made through physical transport of notes and bills using bus and taxi drivers.

Under the M-PESA system, agents manage the transfers with mobile text messaging, allowing simple and immediate real-time transfer. The result is that M-PESA has rapidly become the most trusted form of payment in Kenya and the mobile operator, Safaricom, is now the largest financial operator in the country.

By 2010, M-PESA had attracted 9.5 million customers, increasing to 17 million by 2013, of which over 10 million make at least one transaction per month. Add on the other mobile money operators in Kenya and a large part of the country's GDP is now transacted over the mobile network. For example, 142 billion Kenyan Shillings (ksh) were transacted in the month of April 2013, or $1.6 billion, which would translate into around $20 billion a year processed via the Kenyan mobile payments network in 2013. Kenya's

The increasing use of mobile

GDP was $37.23 billion in 2012, and so it is clear how significant a proportion of the economy is now dependent upon mobile money.

In addition, M-PESA converted many unbanked into banked users, with around 2.5 million bank customers in Kenya when the system launched in 2007 increasing to over 10 million today.

This success has led to many other mobile models from Standard Chartered Bank's partnership with MasterCard and airtel Africa to Easypaisa that works with Orange in West Africa, Safaricom in Kenya and Tigo in other parts of Africa.

Sophisticated mobile services

AS YOU JUMP FROM THE BASIC SMS TELEPHONE MESSAGE through WAP to the iPhone and Android, you find more and more interesting uses of mobile and mobile payments and banking. It is expected that the emerging economies will be at the same level of smartphone usage by 2015 as developed economies are today.

The smartphone enables touch-based access to the mobile Internet, and has completely changed the thinking about how mobile telephony works. This is demonstrated by the sudden rise of apps[2] and the implications of this.

What the app represents is a two-fold change to the way we think about the world and, in turn, the world of money.

First, the app makes everything simplified and highly functional. It overcomes the fat finger syndrome. It can be used equally well by a seven year-old or a 70 year-oldy, without any training and just

---

2   Apple launched the iTunes app store on July 11th 2008 with 500 apps. A year later, there were 55,000 apps and over a billion downloads. In an announcement in June 2011, Apple said they had reached over 14 billion downloads with 425,000 apps and then, by 2013, the store had surpassed 40 billion app downloads with a library of over 800,000 apps. http://en.wikipedia.org/wiki/App_Store

as consistently. In other words, it's foolproof. For complex financial processes, this makes the touch-screen tablet and smartphone ideal. Second, the app breaks down functionality into pieces. As a result, banks can break down processes into bits and make them interchangeable and plug-and-play. You can have a payments app that has nothing to do with a bank app. You can have a balance and alerts app that has no relationship with an account. You can deliver a micropayment and even a nanopayment.

The last point is, in fact, the most important, in that because you can deliver a nanopayment, the world of money changes. This is demonstrated by apps such as Angry Birds which recently celebrated its 250 millionth mobile download[3], and shows that a firm can make millions from charging just 99 cents a download.

In fact, this has led to an explosion of alternative payment options and virtual currencies. For example, in-app gaming on phones and tablet computers is becoming standard, with Apple paying out $5 billion to their app store developers in 2012 and Google paying out a further $900 million.[4]

Given that these billions of dollars are made up of 99 cent downloads and in-app gaming updates of a few cents each via iTunes and the Android stores, it becomes apparent how new forms of virtual currency are in operation. You may disagree, but the aggregation of large amounts of small payments is effectively building a virtual currency system.

---

3   "With 250 Million Downloads Angry Birds Moves Into Magic, Cookbooks, And More", *Techcrunch*, June 2011 http://tcrn.ch/kR2GJY

4   As of summer 2013, Google dominated the smartphone mobile market with 900 million users, while Apple has 600 million iOS, and Microsoft a far third place with an estimated 12 million Windows Phones sold. See http://www.forbes.com/sites/tristanlouis/2013/08/10/how-much-do-average-apps-make/ for more info.

The increasing use of mobile

Other mobile financial services

Camera

ANOTHER PART OF MOBILE worth noting is the use of mobile cameras:

- ❏ CHEQUES – mobile phone camera shots. These are being used to eradicate paper cheque deposits in America, as an image of a cheque is just as acceptable by law as the cheque itself. Therefore, customers can just take a picture of a cheque and send it as a multimedia text message to the bank, and the cheque is deposited. This is one of the most talked-about developments in mobile applications in the USA.
- ❏ COMPLIANCE – banks are under real pressure to be compliant, especially around KYC, so any quirky transaction may result in a request to send a selfie – a picture of yourself.

QR codes

IN ASIA, IMAGES are being used along with QR codes[5] to support completely automated banking. For example, Jibun Bank and eBank in Japan both accept account opening on the basis of just a photograph via mobile of your driving licence. The driving licence is read by a character recognition system and checked with the government's driving database. Providing all is aligned, the account is opened.

More recently, Barclays Bank in the UK introduced QR codes to their P2P payments app, Pingit. The app allows billing companies to send paper payment requests to customers with a QR code and, if the customer uses their smartphone to read the code,

---

5   A Quick Response (QR) code) is a barcode that is readable by camera telephones. The code consists of black modules arranged in a square on a white background and can be linked to text, URL, or other data.

all of the billing information and customer account information is embedded with the code so that the customer simply has to confirm payment. Once this is confirmed, the complete transaction is performed electronically, radically changing the accounts payable and receivable cycle for corporates, as this can now be automated end-to-end.

Mobile contactless payments is another major and growing area of interest for banks. Contactless payments are based upon Near Field Communication (NFC) chips, which are being integrated into nearly all new generation mobile devices.

Apple has intimated that the iPhone will incorporate such functionality and Google estimates that every smartphone will have an NFC chip inside by 2014.

It is in Google's interest to push this capability, of course, thanks to the launch of Google Wallet.

Google Wallet was launched at the end of May 2011 in partnership with Citi and MasterCard. The Wallet is a contactless mobile app that allows MasterCard's PayPass to work with Google Checkout to store credit cards, offers, loyalty cards and gift cards, but without the bulk of a traditional wallet. In addition, smartphones can automatically redeem offers and earn loyalty points synched with Google Offers.

Initially, Google Wallet was only accepted on approved telephones and networks, which limited its success; however, the number of installs of the app surpassed five million in February 2013. Interestingly, Google has now integrated the payments capability into Gmail, so the combination of the Wallet and Gmail should see Google becoming a serious player in the mobile Internet payments revolution.

The increasing use of mobile

The bad news about mobile

As every bank is getting into mobile, there are issues. A good example is the coordinated ZeuS malware attack in Q4 2010, where a web application supposedly from the bank asks the victim to input their mobile phone number. The victim is then asked via text message to install an application onto the phone and the application is used to intercept any text messages the victim sends thereafter.

There is also an increasing number of new man-in-the-middle and mobile malware attacks by the day such as a recent Facebook update about Justin Bieber, which resulted in over 100,000 views in 24 hours with 27% via mobile Facebook. Every viewing downloaded malware.

There is also mobile hijacking. This is where you think you are on your mobile carrier's network but you're not. A cybercriminal places a signal box near to the location of the person they are targeting. The person then sees their mobile signal disappear and come back stronger. What the mobile user does not realise is that their mobile service has been hijacked and all of their texts, apps and downloads are being filtered by the cybercriminals' service. Something that is quite easy to set up and run by any average person with such criminal intent.

Although mobile attacks are increasing daily, banks are viewing this as something that is currently controllable. The question is, with every person on the planet using or accessing a mobile device, is this controllable in the future by a single bank acting on its own?

Making a payment is becoming available to all

This is illustrated particularly well in Somalia, a country which lacks a functioning government but where 34% of adults use mobile money (often to receive remittances from family members abroad).

The fact is, anyone, anywhere can now send and receive money anywhere, anytime.

In Kenya it has created financial inclusion, as discussed earlier, thanks to mobile payments creating credit history and creditworthiness. This revolution is not just confined to emerging or developing economies, but is also occurring in developed economies. For example, the European Union is encouraging financial inclusion, as is the USA.

P2P mobile money makes a difference here, but it's more than this. The real focus of mobile payments and mobile financial developments is all about the war on cash.

Visa, MasterCard and the banks are all trying to encourage the move away from paper payments (cash and cheques) to electronic payments (card and mobile). This is a transformation supported by the mobile revolution, but such a transformation does not take place unless people, process and technology change.

This is the key to the mobile revolution, as what we have here is technology transforming people and processes.

The technology is mobile and contactless, but it's more than this – it's the connected planet.

Until recently, businesses were connected with businesses and governments with governments via mainframe systems.

That changed with the PC, but the PC only connected those with a lifestyle that covered the costs of the technology. That meant only those who could afford one, and limited the world of connections to the developed economies' consumers.

Now, thanks to mobile ubiquity and low cost, everybody is connected wirelessly.

The process changes because it is not just simple transaction engines that are in the hands of individuals, but a whole range of mobile financial services from mobile contactless to mobile proximity, from mobile money transfers to mobile bill payments, from mobile online payments for physical goods to mobile online payments for digital goods, from mobile as a point of purchase to mobile as a point of sale, from mobile as a loyalty programme for coupons and offers to mobile identity and authentication.

The mobile planet is a raft of innovation and change, and it is difficult to keep up with all this innovation and change in our processes because we are hamstrung by our heritage.

We embedded our world in old-style business-to-business systems, and now the consumer-driven world is demanding rapid change to those systems, but they are hard to change.

We are also seeing so many forms of payment – contactless, QR, mobile – that it challenges us to know where to focus and invest. After all, changing processes means changing organisation, products, services, structures and that's costly and upsetting.

First, we need to break the shackles of being hamstrung by heritage. As many people have noticed, the only place we engage with old technology is when we go to work. Now the employees want to bring their own device (BYOD). How dare they, even though enterprises save money on not providing the equipment? BYOD is promptly rejected, especially so by the banks on the grounds of security risks. The digital age will soon solve this. BT, preferred supplier to the US and UK security departments, launched such a solution at SIBOS 2013. But is the enterprise itself fully secure?

Chapter 5: Changing the payments structures

EACH COUNTRY HAS EVOLVED its own payment and clearing systems. Each has their specific rules surrounding corporate governance. The central banks historically had a lock on their own unique payment clearing rules and regulations. These rules are becoming in sync with each other as the world is becoming more interconnected. The level of commerce over the Internet, especially when buying and selling in multiple countries, is growing rapidly.

The EU, with the creation of the euro, planned to harmonise the money being moved between the EU countries. Studies prior to the euro showed fees for international rather than local payments cost often 10 times as much as the domestic charge. Since 2001, a directive (2560/2001) limited bank fees on cross-border transactions to be the same as domestic charges.

The EU then published the Single Euro Payment Area (SEPA) scheme book in 2006, which outlawed float. It is in the banking community's self-interest to keep money in the system, unallocated to any account for as long as possible. This is called float. Float is the money earned by banks on money in transit between accounts. Banks then invest this 'free' money into interest- or fee-earning instruments (investment vehicles), and retain the profit this generates.

The money is free because the bank does not have to pay interest on the money, and so the bank's margin on float is 100%.

Float has to be balanced by customer service and the laws relevant to the timeframe to process, and the EU has defined this timeframe as five working days. After that time, customers can claim for compensation against the originating bank. This is a huge change as, previously, a cross-border payment could take a month. The length of time a bank can keep funds in transit often becomes a political issue in many countries.

In the UK, for example, the banks use an Automated Clearing House (ACH) called BACS (Bank Automated Clearing System) to move money, but it takes three working days. This timeframe was taken as a standard, as it matched the time it would take a cheque to clear prior to automation. However, technically payments can now be processed instantaneously and, with regulations requiring faster speed, the UK is now leading the way.

Faster Payments, the UK scheme, can be used to move money in seconds provided the bank sending the money and the bank receiving it have installed the required software and procedures into their IT environment. This is a critical new service, and has moved the UK towards real-time payments, with Sweden, Poland, Australia and other countries following this lead.

The digital age requires immediacy.

Which payment options wll make it in the digital age?

Let's look at the payment options and see the chances of making it into this era:

Cheque	Minimum 5 days in clearing – not for the digital age
Automatic Clearing House	
One-off payment	24 hours – not really good enough
Direct Debit[1]	3 days - no chance, 3 business days is a lifetime
Faster Payments (UK)	within 15 minutes – ideal
Bank-to-Bank	24 hours – not really good enough

SWIFT – transporter	within 15 minutes – ideal
PayPal	same day, anywhere in the world – ideal
SEPA Direct Debits (SDD)[2]	3 days – better within 24 hours
FX Spot	48 hours – no chance
International Payments	5 days – no chance

1. It is estimated that up to 10% of the existing Mandates are for services and products not now required by the payer.
2. The EU has set the end date for SEPA to be in place by February 2014. The EU's legally binding date is to ensure EU wide rules to make the payment service fair, eliminate hidden charges and accelerate transfers and in doing so could save €123 billion within 6 years. In theory all the banks will finally be able to receive and send SEPA Direct Debits on that date. A lot depends on the companies being ready. The latest number of companies expecting to comply by that date is 30%.

BEING DIGITAL WILL HASTEN THE DECLINE of any payment method that is not at a maximum 24 hours. Near real time is the new standard. The FX standard Spot settlement is two days. The reason for this was that it took a day to send a cable to the US and a day to receive a cable back from the US. Many trading positions are called the Cable Desk.

The entire trading infrastructure has been built on two-day settlement. The Internet FX traders often ignore the settlement timeframes and simply provide the customer with same day settlement. The company then takes the risk rather than upset the customer. PayPal provides a similar service with its customers' international payments, with delivery in less than 24 hours.

The digital age will also see off the epayment and early mpayment services the banks provided to the corporate markets. The epayment services are custom-built by each bank with often limited options to use another bank. The digital age gives the corporates the choice of any bank. The issue for the corporates is to re-build their own internal systems that 'built in' the bank's epayment unique bank message structures.

Changing the payments structures

The use of ebanking started in the 1980s with priority systems from each of the major corporate banks. These systems were expensive to build and maintain but they met the corporate market need to look after their money.

Historically, the larger banks held the skills to create their own ebanking and cash management software. Now many software companies are using XML (Extensible Mark-up Language), an open software computer language, and can provide these services on a licence or as an Application Service Provider (ASP) using internal or cloud technology. Cloud technology is starting to absorb ASPs, making the application and infrastructure a single purchase. The Internet has also replaced the need for propriety communication networks, which the major banks used as their competitive edge in the 1980s and 1990s, by offering cash management business. The sheer scale of telecommunication growth and usage has enabled any bank to enter the digital banking market.

The final frontier for payments – same day international payments

THE DIGITAL AGE will see international and domestic payments on the same timescale, often in near real time. The Nordic countries already do this today.

Digital banking will also change 'correspondent banking' beyond recognition. Correspondent banking relationships started when the domestic bank wanted to deliver services to customers when that bank had no physical presence. Banks went to correspondent banks when their customer left the bank's geographic reach. Digital removes this need.

The traditional payment banking infrastructure for cross-border payments involves four banks. These are the bank of the payer, its correspondent bank and the bank of the biller and its correspondent bank. With the mobile phone, the customer is never out of reach, the need for physical presence is greatly diminished and payments need involvement of only two banks. So what will the two redundant banks do now?

Many banks that grew internationally opened physical branches. These international branches were often 'mini-me's' of the domestic bank; now the digital scythe is coming, as many services can be offered from one location.

The correspondent banking system is well established worldwide and the infrastructure of many institutions often permits only payment routing over SWIFT. SWIFT was established to bring standard formats and telecommunication to international payments. PayPal has accounts worldwide and makes millions of payments over the banking infrastructure using a warehouse account. By using a warehouse account held by PayPal and similar companies, the money is first lodged in that account coming out of a bank account or off a credit card, enabling prompt and convenient payments anywhere.

Payment fees – who pays?

BANKS HAVE CHARGED FEES for payments either explicitly – a known amount per each kind of transaction, or implicitly, depending on the overall relationship the bank. The more banking products and services taken by the client, the lower the fees charged. Each bank sets its own rates. In Europe, payment fees for the euro are limited by law to encourage market price stability and prevent excess fees being charged.

Incorrespondent banking payment, it is often commonplace for each bank in the chain to take a fee. This has also been addressed by the EU, which allows the originator (payer) to pay all the fees along the payment path instead of the beneficiary receiving less money due to the fees.

Transactions for warehousing are usually based as a percentage of the amount transferred into the warehouse that can vary from 2.5%

to 5%. Once the money is in the warehouse, additional explicit fees for transferring the money to other accounts can be levied. PayPal, for example, charges a percentage of the amount ranging from 3.4% for £1,500 per month dropping down to 1.4% for over £55,000 per month plus a transaction charge of £0.2.

Most people expect to pay for a payment; the question is how much. This is down to the individual bank, and in the EU, the law, which states a maximum of €1 per transaction in euros. The new digital companies are using the card exchange schemes as a way to charge 1.5 to 3% of the transaction.

Digital brings together the best blend of domestic and international cash management into global cash management. Global cash management used to be the preserve of the the top 10,000 corporates. It was here the most sophisticated of cash management services were developed and applied. With digitalisation, this sophistication can now be offered beyond the top tier of companies. An early example of this move from elite to commonplace is the sweep account. Designed initially for the larger corporates to move money automatically into a money-earning account, it then became widespread across the banking spectrum.

The other major changes that are happening alongside the Internet are the increasing advance of XML, the formation of global standards for banking transactions and the ongoing advance in nternet security.

Chapter 6: Foreign exchange markets move first into the digital age

As THE DEALING ROOMS were already in real time, the advent of the digital era was taken in its stride. Foreign exchange (FX) has utilised the Internet to become fully transparent and open to all. Citi in the 1990s was not only #1 in the FX business it, through a subsidiary, was the first to offer currency rates that could be bought and sold over the Internet. Citi also created EBS, which allows dealers to post currency prices by lot size across major currencies that could be bought or sold. This early move to digital was not something the bank's hierarchy could endorse over the incumbent businesses being affected. The subsidiary was sold and Citi is no longer #1 in FX.

The issue was that the FX business was enormously profitable, but it became clear in the late 1980s that this was unsustainable. Citi had put in dealing room technology, a worldwide communication network and had the credit appetite to underpin the market. The issue was that the money was so good that it had to be protected, regardless of what was happening in the market and irrespective of what the customer wanted. This is very similar to what is happening to digital at many banks right now.

Before installing EBS, the voice brokers were asked if they would like to be part of this. The reason was EBS removed the need for a voice broker on the major currencies. The voice brokers also made a lot of money helping the major banks balance their positions across their own international branches. The overriding view was that technology would never catch on and they would continue to prosper. The overriding view of the brokers was that the Internet was for leisure, not business. The voice brokers were decimated by EBS almost immediately.

The key to buying and selling FX is either cash in the account or a credit line limit from the bank. Many of the online FX dealers take the money straight away. The use of the credit line, especially for corporates or high net worth individuals, is to cover settlement. A trade in Spot currency takes two days to settle. In that time, the currency purchase can increase or decrease in value. The risk is that one leg of the currency may not occur on time, or if a bank fails, not at all. The infrastructure is based on two days and so the new real-time currency requirements are introducing new demands. These demands are being handled by running many closes of days in one day. That is, every day now has up to 10 computer days.

Clients are often given prices off a pricing feed. A pricing feed is the capture of the price of a currency in real time and offered out to a client electronically. The major price feeds of the full range of currencies – major, secondary and tertiary – are provided by the larger global banks. The vendors such as Reuters and Bloomberg provide rate information on-screen. The global bank price feeds are used as the wholesale price. Vendors and the smaller banks take the major price feeds and offer them to their clients.

The banks provide the currency prices with a built-in spread. The retailer aggregates the clients' amounts and when an amount by currency has been reached and the credit lines established with the global bank, a trade is executed. The vendor or bank then gives the trade to the global bank to change that currency back into the home currency. This gives the retailer virtually no exposure to the market. The positions being created by the aggregation of the smaller trades are such that they are not exposing the retailer to major market shifts in the wrong direction.

To begin automated trading of FX, the retailer needs the money being traded to be in their own account. Similar to the warehouse payment movement, money is transferred into an account under the FX providers' authority. Once money has been put into that account, the client is ready to buy and sell currencies.

The corporate can see and trade (execute) on a moving market FX rate (the rate is usually held for 30 seconds by the trading party) from an online FX service. These services are readily available through banks and non-banks. Often, local banks can brand an FX service provided by a major financial institution, with the market risk covered by that institution. The corporate now has an exposure in another currency (if it is not a straight one-off payment). All this is possible because the FX market embraced the digital age.

Effective management of foreign currency risk stabilises the company's performance. Currency risk management programmes can structure a cost-effective currency management strategy to match the risk exposure. A comprehensive set of currency management services including spot, forward and options, are available.

For example, a derivative in the form of an option can be used to hedge a currency position by buying a future value for that currency. The option represents a contract between two parties. To buy at a specific price is called a 'call'. The right to sell at a specific price is called a 'put'. Options have an expiration date and come in two forms; American style, which can be exercised at any time; or European, which can only be exercised at expiration.

Precisely when the corporate buys and sells the FX can be a significant business driver. A classic tale of such activity was when Lord Cowley, then head of car production in the UK, after being told car manufacturing had lost £300 million but the treasury area made £50 million on FX noted: "Let's get rid of the car manufacturing".

Most banks offer an FX 'branch' rate usually in amounts up to $50,000 for transactions from one currency to another. The rate is changed once a day and contains a substantial spread, often 6% (depending on currency) either side of the midpoint market rate, plus a transaction fee. It is cheaper and often easier to use online services but treasuries are often undermanned and staying with the proven is often the way.

FX rates for a pair of currencies are now more easily established from the Internet. Some corporate internal policies insist on three quotes. The main corporate function is to make payments and these tend to occur at the end of the month.

Currency exposure management

COVERING FUTURE CURRENCY MOVEMENTS is essential. As soon as the order is placed in another country, book an FX forward to cover the amount payable or due. For example, for buying steel from the US for delivery in the UK in six months' time, a six-month forward that sets the FX exchange rate can be bought. In six months' time, regardless of USD and GBP exchange rate fluctuations, the margins on the steel contract remain intact. The cost of the steel is the same as it was six months ago. If the price of the pound had strengthened, the steel would be cheaper, and similarly if the US dollar had strengthened, the steel would cost more.

Once a company has two or more bank accounts in different currencies, currency exposure is created, as any two currencies are rarely static over their values of exchange at any particular time. The market is continuing its own valuation of a particular currency and changes can be dramatic. Every company has a statuary currency – its home currency – in which accounting and audit reports have to be filed. Consequently, accounts in a currency other than the statuary currency have to be shown usually under the term 'currency exposure'.

Currency exposure can affect the profit and loss account (especially if there is a major currency movement either up or down), the value of international assets, the risk in long-term international contracts, sales revenues, and borrowings in international currencies, and the timing of payments being made and received are all critical.

Here a rolling schedule of incoming fees and expenses in various currencies can show where the market risk is at its highest. Trading with a country that has consistently high inflation often results in lower net revenues coming out of that country due to the difference in exchange rates. A country such as Australia, dealing with the UK, has seen its currency rise in terms of what the UK pound (GBP) can buy. Hence if income is coming out of Australia, the revenue is increased, but if costs are being incurred they will be more expensive.

Margin (or increased leverage)

THE RETAILER CAN PROVIDE THE CLIENT with a margin account that enables the client to buy more currency than he has in the account. The reason for this is that the retailer is looking at the movement the particular currency is making in a day. It is rare for a major or secondary currency to drop or raise by more than 10% in any one day. Consequently, the retailer with the authorisation over the account can sell out of the currency when required, should the money in the account not be sufficient to pay if the rates move downwards.

Depending on the retailer, the client's profile and the past practices, the following margins can be issued for a major currency. Assuming 10% daily historic movement, then the retailer could give a margin of 10 to one. That means the client can buy 10 times the amount in the account. By putting up £1,000 the client can buy £10,000. The retailer probably has a stop-loss order.

Living in a multi-currency world

A CLIENT COULD have an account in pounds (GBP) and has bought euro (EUR) and sold US dollars (USD). The client could be looking for short-term gains and closing out the position at the end of the day. In this way, no market risk is taken overnight.

Traditionally, companies have kept their global cash surpluses in US dollars, Euros or GBP. This process is now undergoing a review, especially with those companies seeing their Asian operations growing and becoming a larger part of their revenues and expenses.

Asian businesses have historically maintained far more currencies than those in the US and Europe. The reasons are often high interest rates in a particular currency, and high inflation that in turn creates volatility of that currency. Most Asian bank account clients use more than 5 currencies, whereas in the US and Europe, the majority stay with one currency.

The Chinese authorities are encouraging the RMB (renminbi) to be used as a settlement currency for international trades. Hong Kong is a major beneficiary of this policy, with its deposits in RMB doubling in the last few years, and it has the largest offshore deposits of RMB in the world.

The European Central Bank report shows the RMB has become a key driver of currency movements in Asia since the mid-2000s, and even more so since the financial crisis. The results suggest China's dominance in the region, and the role of the RMB in the International Monetary System, make it already a tri-polar currency. That is, the US dollar, euro and RMB are now equally acceptable in the market.

Without digital technology the FX market could not be accessed by everyone at any time. Citi showed what can happen to a bank if the digital movement disrupts the status quo. Imagine where Citi would be if it had gone with the subsidiary and applied it across the entire bank.

Digital means intraday banking

THE NEED TO STAY LIQUID is paramount. Many banks during these financially difficult times are selling non-core assets to meet their regulatory obligations. Banks and corporates now need to manage intraday liquidity as they are receiving and paying out large amounts of money throughout the day.

While the trading floor often knows its actual cash position across assets within seconds, banks and corporates, because of their size and complexity, often do not. Most banks and companies still rely on end-of-month reporting for their liquidity. Given the increase in money being moved in near real-time, the need to know the intraday position becomes vital.

The big debate in the industry is whether to charge for intraday overdrafts. The US Fed was the first to charge for intraday credits with a penalty structure for its CHIPS product. Many trading rooms have a basis point charge for the cost of money should a trade require credit for the day. A trader making 10 basis points is outweighed by the cost of the intraday overdraft.

Each institution has a different cost for intraday credit. So the days of the colleague bank paying out the money first on a syndicated loan are gone. Before the lead bank pays the loan, it will ensure everyone who has a slice of that loan pays them first.

So what happens when a payment is requested though Faster Payments and the covering payment is in the network? All banks have ongoing experience of intraday overdrafts. Today, they cover them themselves which is expensive. The collateral for the cover is cash-like instruments, which are premium priced. So who pays?

One bank has looked at 'Dollar Seconds'; when you are overdrawn, you are charged by the second. The complexity and room for unintended consequences are high. The Federal Reserve in

1994 introduced daylight overdrafts fees. The move changed behaviour, which was the intention, but the risk profile changed also as banks were releasing money later in the day. Now new rules have been implemented that 50% of the money on account leaves by noon. The risk created was that all payments occur in the last minute of the day.

Alerts to intraday liquidity will become essential

INTRADAY LIQUIDITY IS NOW A MAJOR CONCERN for everyone. Many check the balance on their bank account on a daily basis. Soon the balance will be checked prior to the payment being made. Also alerts will be set should transactions over a certain amount arrive or leave the account. This becomes easy once everyone in the chain is digital. Companies, including Board members, can see the liquidity reports.

Companies always have to balance the collection of money in before the requirement to pay it out. Most want to be paid immediately and then enjoy as much time as possible before paying out. This especially so in the interbank market, and this resulted in the Central Bank becoming piggy-in-the-middle when the banks did not trust each other in the credit crisis. "Hold the money" is the banks' motto.

Degrees of liquidity

ASSETS CAN NOW BE GRADED by degrees of liquidity. This is the time it takes for the assets to be sold and converted to cash. The Internet in the UK provides the cost of each house and its potential value. The mystique of the pricing by real estate agents has gone. The Internet has made pricing of items easy to research.

Banks often left credit with other banks before the financial crisis. Now they tend to keep it with the central bank. This consolidation of credit to the lender of last resort has left many banking systems short of credit flexibility.

The digital supply chain

THE INTERNET MAKES BUYING GOODS anywhere easy. The Internet has no borders. Consumers want to buy the product they need quickly and easily and pay for it with the same simplicity and immediacy. The issues come in the wholesale space. Here, industrial sourcing of parts, and its international nature, falls back onto the traditional supply side.

The basis of today's trade finance started with camels crossing the deserts in 5,000 BC. To ensure the risk was contained, as they disappeared over the horizon, a raft of familiar instruments were created to protect the parties involved in the goods and finances. If they had smartphones back then, these instruments would probably not exist:

❑ Letter of Credit;

❑ Bills of Exchange;

❑ Promissory Notes;

❑ Standby Letters of Credit ;

❑ Letter of Guarantee.

We have enjoyed the humour of slowness of the various legs of international trade. Over 50% of goods, by value, are delivered by air and arrive often days before the cash is moved. The supply chain is riddled with breakpoints. The issue is the paperwork and the flow from accounts receivable to accounts payable of the various parties. Nearly 60% of accounts payables are paper-based.

There is also the recognition among buyers that they cannot afford to have goods in transit or inventory for any length of time. In the UK, for example, the entire country's petrol (gas) supply at service stations is about seven days. The cost of inventory is not economic, as duty has to be paid immediately (duties in the UK are over 80% of the price of the fuel).

'Just in time' marketing has been with us for a while but this is stymied by the supply chain. At each stage of the chain, the person who owns the goods wants to be paid. Similarly, so does the carrier, and any trade barrier fees need to be paid or the goods stay where they are until money or a guarantee to pay arrives.

SWIFT estimated there is €3.67 trillion in trapped capital in the corporate supply chain. That is money a company may not know about and therefore is unable to re-invest or deploy to another part of the business. In addition, since 2008, this has increased by €183.5 billion as debtors have slowed payments by 4.3 days of sale. That means that, as the digital age is speeding up delivery, the pre-Internet world of the supply chain is slowing things down.

As most economists will tell you, money being circulated faster means better economic growth. An example of this is that, for every £1 spent on the building of Crossrail, the largest civil project in Europe, £2.84 is generated. The trillions trapped in an arcane system is crying out for release and only new thinking tied to the digital age will release it. Time, as the song goes, to put the camel to bed.

Financial supply chain innovation has tended to be bi-lateral, with a bank creating a specific solution for a customer without integrating the customer's trading partners. This is especially so on the payable side where the buyer and the bank control the cash flows with relative ease. This approach does not permit the use of others in the supply chain. SWIFT's Trade Services Unit (TSU) has been struggling to find transactions, the volumes are very low, but it involves multiple parties. There has also been a move away from the older style instruments, especially Letters of Credit, and now 80% of global trade is now conducted on open account. This move permits the use of digital activities to increase.

The SWIFT's TSU is a centralised matching and workflow platform available to the SWIFT community. The platform provides comparison of data taken from underlying corporate purchase agreements and related documents. The TSU supports the exchange of the new 'Bank Payment Obligation' (BPO). BPOs are an irrevocable conditional obligation from one bank to pay another bank, subject to the presentation of compliant data in the TSU. The 'Notice of Intent to Pay' message is an additional information message indicating the corporate intent to pay another corporate. Together, these two features provide a strong backbone for banks to offer alternative forms of financing, including pre-shipment, post-shipment and reverse factoring. But only if they are part of SWIFT's propriety community.

The digital age is sweeping away closed-end user groups such as the supply chain when the incumbents do not or cannot provide the level of services needed now. The trade unions of yore, protected workers' rights; now largely these rights are protected by law so the trade unions must reinvent themselves. Similarly the TSU at SWIFT needs a healthcheck on surviving outside the digital age.

The common denominator in the supply chain is the invoice. Everyone in the chain issues an invoice and then monitors it through the accounts payable or accounts receivable departments or in 'Tub files'. Yes, some companies still use Tub files, which like Telexes are part of the pre-Internet period.

For the digital age – the e-invoice

THE EU IS TRYING DESPERATELY HARD to introduce e-invoicing. E-invoicing is the electronic transfer of invoice information (data on billing and payment) between business partners (supplier and buyer). It is part of the supply chain linking internal processes of the corporate into the payment systems. The EU book on each country's criteria for invoices runs over 100 pages.

The EU estimates that replacing paper invoices with electronically generated invoices could save €64.5 billion per year for businesses. The problem has been the legal and infrastructure issues in each country and within the corporate internal environments. The EU has a goal to make e-invoicing the norm rather than the largely manual process it is today through the use of a legal framework and standards including ISO 20022.

In Europe, while e-invoicing adoption is growing, it still has a long way to go in terms of dominant market share. In most of Europe 15% of all invoices are being issued electronically, the remaining 85% manually.

Finland is the #1 country for e-invoicing. The State Treasury of Finland offers companies that do not yet have an e-invoicing capability a free-of-charge service for sending their invoices in electronic format. So even the smallest suppliers can shift to e-invoicing and start to benefit from optimised cash flow, better working capital management and lower processing costs.

Finland estimates an incoming paper invoice costs €30–50 for the receiving company. Companies can reduce these costs to €10 by semi-automating the invoice process, and reduce further down to just €1 by fully automating the process.

It is not surprising then that the first major provider of mobile phones, Nokia, is a Finnish company. Finland adopted digital technology readily. One of the reasons for its ready acceptance is the weather. Temperatures of −50 degrees in winter are make day-to-day commerce difficult. As the adage goes, "necessity is the mother of invention", so Finland was an early adopter of digital. Interestingly, they found that the carbon footprint for an outgoing e-invoice is less than half of that of an outgoing paper invoice.

Finland is not the only government driving the adoption of e-invoicing. In Europe, there are similar initiatives ongoing in Sweden and Denmark. Outside of the EU, in Mexico it is now mandatory to issue invoices electronically if the invoice sum is higher than 2,000 pesos, and, since 1 January 2011, businesses are required to perform regulatory compliance checks for senders' and receivers' invoices. This can only be done efficiently through digital practices.

In the USA, e-invoicing follows a set of procedures established by the National Automated Clearing Houses Association (NACHA). The standard formats and intra-operability rules allow banks to offer a multibank option for businesses. The ACH payment message formats allow banks, billers and e-billing providers to standardise transactions. Companies can now present e-invoices in standard format through multiple banks. Clients can view and pay their bills though their own online banking site via PC or mobile phone. Kind of getting there digitally, but the emphasis is still on the client making a special journey to the bank site.

The goal with e-invoicing is to reduce the paperwork associated with accounts receivables and payables. Consumers in the USA pay 50% of the bills they receive electronically each month. The consumers use the Internet to locate their bill and then pay it on line. Basically the billers post their invoices on line and e-mail or text the buyers. The buyers log on through a secure web site and pay their invoice. The buyer often has to go to many web sites to pay for many different bills which starts to defeat the purpose of being digital.

The US pioneered e-invoicing through the Financial Electronic Data Interchange (FEDI). However the required technologies needed at that time were only financially viable for the larger corporates. The key drivers of FEDI were GE and Ford. All suppliers to these companies were required to send in the invoices digitally. Paper is rife in the US, with the postal system delivering an annual 26 billion invoices and payments in 2011. The cost of printing, handling and mailing invoices to consumers is estimated at $21 billion a year.

The benefit of electronic invoice processing is a significant reduction in invoicing costs:

The number of invoicing errors is reduced because the invoice information is delivered to the customer after being fully checked automatically;

❑ Delivery of e-invoices to customers is immediate;

❏ The invoice can be processed automatically with little or no manual rekeying of data;

❏ Increasingly transparent processes.

Help is at hand for the e-invoice: there is a legal commitment on the banking community to support banking specific BIC (Bank Identity Codes) and IBAN (International Banking Account Numbers). These are unique strings of alphanumeric characters that identify each and every bank account. Ultimately these strings need to pass through the many underlying clearing systems to move money from one account to another.

To gain these efficiencies for the corporate, there is a need to add BIC and IBAN numbers to invoices to their clients. More information is needed on the invoice, especially to international clients. As BIC and IBAN are long streams of numbers, the chances of them being accurately transcribed manually are low. Hence the corporates and the banks have to embrace digital technology to invoice and be paid quickly.

This removes tremendous inefficiency from the international banking system, which is good for commerce and a mixed blessing for the banks. The banks want to remove the inefficiencies – repairing international payments, which is a small industry in itself, for which the banks incur the expense – but removing the inefficiencies of the payment processes also costs the banks money in reduced float. Banks in general profit from inefficiencies, presenting a natural barrier to the rise of speed and improvements and hindering the move to digital.

The right data creates inherent benefits

THE SUPPLY CHAIN uses many bank products and includes payments, cash management, trade finance, foreign exchange and information. Information will drive many things.

The introduction of new XML-based payment initiation messages, combined with the associated advices and bank statements, begins to address components of corporate and bank messaging. The XML approach helps eliminate the use of proprietary bank standards and provides the necessary data consistency for banks to provide debit advice and financial statements electronically.

The next step is the corporates themselves. 50% of them use a methodology of paper and spreadsheets; 20% use internal ERP systems; 17% use a module added on to the ERP system or a third-party offering, with 3% outsourcing the process. The remainder is a mixture of standalone workstations, software designed for their specific industry and a wide variety of packaged software. Most predate the smartphone. The saying: "When I go to work I go back in history" is very apt.

The best practices of the top 20% of corporate treasury performers have gone digital and they are:

❏ 1.4 times more likely to use electronic payments;

❏ 2.6 times more likely to have automated alerts to notify exceeding pre-defined thresholds;

❏ 3.8 times more likely to establish pre-negotiated, fixed early discounts and meet them.

The supply chain for larger companies represents a challenge for all parties to make it as liquid as possible. To do so, the supply chain has to move from physical to electronic. The top performing treasuries have shown this can be done digitally.

The current complexity of the landscape creates difficulties in achieving standard workflow control. The ongoing manual dependencies hamper the complete chain, resulting in an ineffective cash management position. From a cash management perspective, increasing digitalisation can facilitate improvements in operational efficiencies, achievement of better STP and the optimisation of corporate cash. In addition the corporate can achieve further operational efficiencies through the elimination of the manual reconciliation process. Ultimately the reconciliation process should occur daily to gain best efficiencies.

Funding outside the banks

THE SUPPLY CHAIN for an investment grade company can use the advantage of their credit rating to support their clients. The really large companies with many suppliers can use their often-excellent credit rating to facilitate invoice financing for their suppliers from their banks, or in the future, the financial digital market.

As the products are being made the money involved in the manufacturing and transportation of them follows them. Each party in the chain requires payment or guarantee of payment before moving the products and the digital market place can play its part.

Invoice financing can be open to all

IN TIMES OF TIGHT CREDIT, the use of invoice financing in all its forms increases, especially when banks do not want additional credit lines on their books. Invoice financing is riskier and labour intensive as each invoice has to be verified, so the interest rates are higher than a credit line, reflecting the additional administration and the risk in the process.

Corporations with good credit ratings tend to pay all in 2–3% per month on the money extended. In addition to the credit reports, credit assessment Personnel Guarantees (PGs) are often required.

Invoice financing comes in a variety of forms: the two most common are Invoice factoring and invoice discounting. The difference in invoice discounting is that the financial company is kept confidential and the suppler (biller) performs the back office functions.

A bank (or finance company) takes a view on the invoices being generated by the supplier. These are checked for validity – each and every time – and often who the customer is and their credit rating. The invoice itself often has to be paid by a larger corporate that often has a better credit record than the supplier and guarantees to pay the invoice usually after 120 days.

The amounts requested are compared against the supplier's credit rating and within the agreed to conditions at account set-up. The bank issues a credit to a level somewhat below the value of the invoice. For example a £10,000 invoice often is reduced by 20% and the £8,000 paid to the supplier. The money is advanced to the company and offset as money comes in from the invoices. Often the money coming into the bank is kept in a separate account, reconciled with the invoices outstanding and then passed onto the client's account.

Factoring charges a percentage of the invoice and/or interest is charged on monies outstanding. The reason it works is that the bank provides the money to the company while the suppler waits for the invoice to be paid. Payment is most often set at 30 days contractually, but often this is stretched to 60, 90 or 120 days.

The EU and other governments are trying to establish standards for the timeframe between invoice and payment. The goal is to have a 30-day standard. The UK government has key performance indicators (KPIs) for many of its services; the NHS (National Health Service) KPI is to pay invoices in 30 days. The current level is around 80% that are paid in that time. This is good, given the level of manual activities of the existing paper flow.

Many companies offer a discount on their charges if the payment is received within an agreed timeframe. This can be a discount from 2–5% if payment is received in 10 days. Unfortunately, the 2%, about 72% APR per year, is routinely missed. For example, there is a US national discount retailer which routinely pays its manufacturers in 56 days, despite being offered 2% discount for payment in 10 days. The system is pre-Internet and cannot cope. The digital age will see that 2% being taken easily.

By taking the 2%, companies reduce the cost of financing their working capital needs and the supplier has reduced DSO (Days Outstanding). DSOs are basically interest-free loans given, often reluctantly, by the suppliers.

The supply chain is full of potential to release money for the parties involved. We can do this by breaking the money pots into three components:

- **DAYS SALES OUTSTANDING (DSO)** – the time to collect payment once the service has been sold
- **DAYS INVENTORY OUTSTANDING (DIO)** – the time it takes to replenish the initial inventory
- **DAYS PAYABLE OUTSTANDING (DPO)** – the time it takes to pay for the goods sold

While all three can be influenced by the corporate, only DPO is fully controlled by the corporate. The goal is simply to reduced DSO and DIO by a third.

Those companies who had automated accounts payable and accounts received noted:

- Processing of DSO dropped to 21 days versus the industry average of 53 days – a digital improvement of 40%;
- Cash flow forecast was 84% accurate as compared to the industry average of 49% – a digital improvement of 100%.

DSO is often noted in credit reports as an indicator of the credit health of a company; a high level of old receivables outstanding viewed against the normal for that industry's sector could flag a concern. In addition, the older the receivables, the less likely they will be fully collected. Again, with digital technology the pre-Internet normal of 30–60–90–120 day catchments can be reduced to the profile of the customer. A good payer with an amount outstanding for 90 days is probably better than one with a poor record at 30 days.

Some Boards and/or management teams focus on DSO numbers while others do not. One example of a change in management was in the international software sector. Misys (then a UK FT250 company) had less than 40 days outstanding in 2005 and by 2011 this had grown to 86 days. The company mantra in the early 2000s was 'cash is king' and any receivables over 60 days were actively addressed and any receivable hitting 120 days was written off against the business.

DIO is the working capital tied up in the inventory pending sales. The longer it takes to replenish the inventory, the longer it costs to fund. This is particularly important in areas where tax and tolls are taken at the point of acceptance.

Days Working Capital equals: Accounts Receivable added to inventory and then taking off the Accounts Payable. This total is then divided by the total per day. (DWC = (A/R + Inventory – A/P)/ (Total Revenue/365)) and shows how many days it takes a company to convert its working capital into revenue. The faster it does this, the better the cash flow and efficiency within the company of collecting money. The shorter the working capital ratio the better is the timeliness of the cash flow. Since 2005 there has been little improvement, staying at around 44 days.

Reconciliation

THE ISSUE IS THE HISTORIC INFORMATION STRUCTURE that surrounds the early days of the ACHs and the early punch cards. The early computer punch cards were 3x7 and with fixed fields. The early computer screens were laid out in a similar format which evolved into banks' core systems being built at that time and which are still with us. Hence, information is in fixed fields and difficult to interact with unless you are a COBOL programmer.

Many an ACH payment has limited remittance information that can be forwarded to the beneficiary due to the 12-character remittance advice field started when giro payments begin. A beneficiary receiving the limited remittance information cannot apply the funds automatically without first finding the customer and then the invoice being credited.

These pre-Internet barriers are overcome by having reconciliation take place automatically. The process should highlight only the exceptions (usually less than 5% of the volume but often representing 80% of the manual cost) and suggest matches through business intelligence built into the process.

Foreign exchange markets move first into the digital age

The automation reduces the cost of administration and customer service is much improved. The need for further automation is clearly reflected as only 29% of companies had fully automated the procedure to pay (accounts payable); and 25% of companies had fully automated the order to cash (accounts receivable).

Reconciliation is often one of the last functions to be done on a timely basis as it often compared to looking for needles in haystacks. It is time-consuming, costly and could be a high-risk business, especially as the regulators do not like to see payments not recorded against the right account and for the right amount.

Chapter 7: How does a bank make money out of being social?

FIRST OF ALL, let's define our terms.

Social media is like mainstream media but user-generated. Its blogging instead of newspapers; YouTube instead of TV; and podcasting instead of radio all done anywhere. It's media, but social media.

Social networking is like real-world networking, but online. It's where people date, talk, meet, relate and even work together, on occasion. It's networking, but socially and remotely.

Social banking is providing bank services but, in a similar fashion, through social methods. It is the process of personalising banking and enabling people and businesses to see the people and businesses they are saving and investing with. It's Zopa and Prosper, SmartyPig and Mint, and many more.

Social money is the way to make payments peer-to-peer, and has already taken off big time with PayPal and Square, but now has many other pretenders, especially in the mobile space.

On that last point, please note that all of these services are based upon Internet Protocols (IP) as the platform, but may be delivered through any device including TV, PC, tablet or mobile.

Social media replaces traditional media as a source of news, views and entertainment.

This does not mean that the TV has been replaced by the Internet as the primary entertainment source but, for a large range of demographics, this may be the case.

Meanwhile, how many of you pick up a paper every morning to read?

How many of you pick up a free newspaper these days?

It's free because print news is worthless. It's out-of-date by the time it hits the streets, so you read it more for the tactile feel, the fact you don't have access to the net or TV, and for the opinions and commentary. More photographs are used and the words are few and bigger as compared to any newspaper before the start of the digital age.

I read what the crowd I follow read.

If my friends say a story is worthwhile, then it's worth reading.

That's social.

So how does a bank make money out of social media?

First, a bank should not try to make money out of social media.

Social media is not there for money-making but for customer engagement, which leads to money making.

And what does it mean for customer engagement?

Well, on the one hand, it understands customer attitudes and ideas.

If customers are out there saying the bank is awful, horrible, difficult, complacent, arrogant, greedy or worse, then it's engaging with those customers to find out why they think the bank is all of those things.

This is what banks like West Coast American bank Wells Fargo and Indian bank ICICI recognise.

When Wells Fargo was asked why they blogged, the bank's answer was simple: "If you're not part of the social world of conversation amongst your customers, then they will talk about you negatively and you have no voice to respond. If you engage in the online conversation, then it becomes far more civilised, interactive and interesting."

Other banks fear such activity, due to the negative feedback they might receive.

Again, Wells Fargo found that there was negative feedback at first but, by having a team monitor their social media 24/7, they always responded to any negativity straight away with a response explaining why it happened that way.

Customers were far more polite and calm when they saw their crude postings garnered a civil reply; it led to being engaged in a conversation. Through conversation, the bank learned a lot more about what frustrated customers. The result is better products and services.

However, it is quite clear that the bank cannot engage in such activity half-heartedly, as you need to be responsive and therefore have people dedicated to social media interaction.

Like a call centre, it's a response team to online questions and issues.

Interestingly, over time, customers often reply to others' crude postings and become the best advocates for the bank. Many other financial institutions have found the same results of using social media, where customers engage and manage other customers' feedback online.

For example, in 2012, ICICI Bank in India launched full service banking in Facebook as an app.

There are over 80 million Facebook users in India and ICICI Bank believed there was an opportunity to engage those users for bank services.

The bank service was launched in February 2012 and achieved a million Facebook Likes in just over ten months. After a year, the bank had achieved over two million Likes, and the Facebook page offers full service banking as well as lots of other services of customer engagement. The most recent innovation of the bank has been a service called iWish, a smart savings tool launched in partnership with the US firm SmartyPig.

SmartyPig provide a social savings tool for banks. The way this works is that you set a goal and then share that savings goal with friends and family to raise funds for your target. Your friends and family can then contribute and support your savings focus, and donate or gift you funds to assist if they feel it is appropriate.

In terms of ICICI's Facebook banking service, the bank provides a secure link from Facebook into the banks services. As Sujit Ganguli, Head of Brand and Corporate Communications at ICICI Bank, said: "Our customers were worried about going into their bank services from Facebook but, as we made clear to them, you leave Facebook and are using our usual secure servers once you accept and install our Facebook app",

In other words, Facebook apps are like mobile apps for banking: it's just another app-based banking service; and ICICI Bank uses it to engage the customer by providing content, care, communication and creativity.

In ICICI's case, they offer gaming and more, to engage the customer and the best part of ICICI's experience is the impact of social media engagement with the bank's clientele.

For example, before using Facebook for customer engagement, 24% of the online mentions of ICICI were negative and only 19% positive. Now, 49% are positive and just 6% negative. That's a massive change and shows that a key part of the use of Facebook banking is for customer servicing, not just engagement.

This is a key factor in using social media: customer engagement.

Therefore, in order to ensure the right image of the bank was presented, the bank sees social media as a key method of moving the right message to the top of the search results, rather than leaving it to negativity from media or anti-bank activists.

For example, the bank could start posting stories that help those people out there who are struggling with finance and show what positive actions to take.

This tends to be the domain of independent blogs and aggregator sites today, although there are some great bank examples in this space, such as Royal Bank of Canada's site aimed at students.

What these sites do is create a relationship with people, through news, views, advice and ideas. That's the point of social media: to create a conversation that leads to a relationship that leads to trust that, eventually, leads to business.

In fact, there are many other new financial service operations emerging using social media leverage across all areas of banking.

This is an unstoppable movement and therefore it is quite surprising that most banks and bankers do not even use Facebook and Twitter. A few do, such as the banks mentioned already like Wells Fargo and ICICI, as well as other such as American Express, Commonwealth Bank of Australia, USAA and First Direct, but the point is that any organisation that ignores the fabric of society – bear in mind over a billion people use Facebook - is missing a trick. Try explaining that to the Board of Directors, who are largely from another era.

This point is being missed by many banks because many banks are firewalled out of these changes.

The workers spend all day working and do not see why social media is relevant because they cannot use it at work and are too tired to spend time at home on a social network, when there are other things to do, like being a dad, mum, husband or wife.

I am sure many banks are changing their policies towards social media and no longer firewalling everyone out ... but there are still far too many that are not and these are the ones that are starting to smell a little bit of nostalgic old worlds.

For example, search YouTube for 'bank' or 'bank advice' and there is hardly anything worthwhile there.

Under the search for 'bank' the first page of results, in order of view count, contains music videos by Lloyd Banks, 50 cent and Azealia Banks, and a very juvenile cartoon of Sponge Bob robbing a bank. Mind you, with almost ten million views, the latter is popular with the social crowd.

That's entertainment!

How does a bank make money out of being social?

What has a social network got to do with banking anyway?

SOCIAL MEDIA EDUCATES, advises and supports, which builds relationships and trust. So do social networks ... sounds like a bank and banking.

The starting point of social finance is paying for things, and that is why the first things financially related that appeared on Facebook are applications such as Pay Me, Spare Change and PayPal, although only PayPal survived the process of evolution. This is because they are the incumbent, and paying between friends is what you focus upon in social networks, not banking.

From a purely social networking viewpoint, banks need to focus upon methods they can use to become friends with their participants in their networks. Friends advise and support, they don't sell, charge and make money out of you. By advising and supporting, banks can build relationships and trust and so, like social media, it will result in advocacy from your 'fans' who then became loyal and easier to do business with.

That is the point: building trust and loyalty.

Facebook and Twitter are not social media but a social platform for customer engagement, as is YouTube, Flickr, Tumblr, Foursquare and more.

Social platforms are like Salesforce.com and the Internet.

They are the building blocks of the new financial firms.

They are not the end goals in and of themselves, but are the enablers towards real social connectivity.

That is why when banks talk about "We've gotta getta blog out there; build a Facebook page; launch a YouTube channel; and engage on Twitter", that is not the answer.

Many firms believe that having a holding Twitter name @ MyBank and Facebook page is all you need.

Some go as far as to populate their pages with links and news.

But you still don't get it if you think that way because these are platforms, not websites.

Facebook and Twitter have hundreds of specialist service providers creating new forms of social engagement from content custody to social marketing management; from social ads to social intelligence; from apps for gaming to apps for sharing to apps for commerce; and more.

A good example is Instagram, purchased by Facebook for $1 billion in April 2012. Instagram is a photo sharing social service, and provides an easy way to share such content via Facebook.

In other words Facebook, like the Internet, is a platform that provides the underpinnings for far more targeted and specialist social connectivity.

Social finance is starting to arrive

IT IS QUITE CLEAR THAT BANKS CANNOT IGNORE these developments, and need to use social media communication capabilities to engage their audience in a conversation that advises, supports and educates potential customers in their financial capabilities.

This advice, support and education can build into a relationship and a trust that might generate future account openings, but that is not the primary intention. The primary intention is to build trust. After all, banks have lost so much trust over the past few years that this alone must be a strong reason for social networks and media to be used as a critical platform for future business.

There are many new financial service operations emerging using social media covering capital markets (Etoro, Stocktwits, etc.), corporate banking (Funding Circle, Kickstarter, Market Invoice, Platform Black, The Receivables Exchange, etc), retail banking (Zopa, Moven, Simple, Bitcoin, etc.), payments (Currency Cloud, Square, mPowa, etc.), and insurance (Friendsurance).

These sites were all newly launched in the past decade and are

firmly based upon proven social models of finance. In general, the services fall into three categories of social finance:
1. Social money and payments;
2. Social lending and saving;
3. Social funding and investing.

Social money and payments

SOCIAL MONEY AND PAYMENTS has been around for a while, with PayPal being a great illustration of the first entrant in 1999. Now 14 years old, PayPal is the incumbent, with new start-ups like Square, mPowa, iZettle and more being seen as the innovators. This is because PayPal capitalised on the first wave of social money – person-to-person (P2P) Internet payments – whilst the latter focus upon the second wave – P2P mobile payments.

This is now a hot space with many different flavours, including:
- Mobile remittances changing the game rapidly, with Kenya's M-PESA copied into multiple geographies.
- Jack Dorsey, the cofounder of Twitter, launched Square which launched in 2009 and has similarly been copied into multiple geographies.
- Google Wallet was an industry milestone when it launched in May 2011
- Virtual currencies like Bitcoin have been nibbling away at the market and are moving mainstream.
- PayPal is finding itself under attack from the card firms such as AMEX, Visa, MasterCard and the banks, with iDEAL and ePayo targeting their schemes for social payments

Square has been the main news story, mainly because of its phenomenal growth and charismatic founder.

Square was created by Jack Dorsey, co-founder of Twitter, in 2009 with the idea of offering a simple, cashless payment system for smartphones. The system works by using a square dongle that fits into the earphone jack of a smartphone. The dongle allows a user to swipe a magnetic credit card stripe and take a payment. The payment is made by the buyer using the app that goes with the dongle to sign the smartphone touchscreen with their finger and enter their email, and that's it. A payment can then be made. In fact, using the Square Wallet app, merchants and customers can then transact wirelessly after that first set-up process.

The idea for Square came from a conversation that Jack Dorsey had with an old friend about the fact that his friend had lost a recent sale – a $2,500 blown-glass bathroom faucet – because his customer could only pay with a credit card. As his friend told this tale to Dorsey – both of them with iPhones pressed to their ears – they realised that a business was literally at hand. Those smartphones had more processing power than entire banks a decade earlier, so why not process credit card payments on telephones?

One of the key attributes as to why Square took off so quickly is that the sign-up process takes minutes. You download an app, put in your name and address, answer three security questions, link your bank account, and that is all there is to it.

Just as Twitter democratised broadcasting, Jack Dorsey's new company would democratise the credit card industry. This resulted in Square processing over $4 million in payments transactions per day by July 2011, just over a year after its launch, and this figure was impressive as it compared to just $3 million in May 2011. By May 2013, Square was processing $15 billion in payments per annum, or $41 million per day, up from $10 billion in November 2012.

This is why Square is being challenged by heavyweight industry incumbent PayPal, who launched PayPal Here in 2013.

PayPal Here is a mobile payment app and card reader for smartphones. It lets users simply and securely accept multiple forms of payment including proximity payments, where the user and the merchant apps use the mobile network to transact the payment with no token exchange involved. In other words, a completely

How does a bank make money out of being social?

frictionless payment, with no action involved other than the use of app-to-app networking.

These players are being challenged by suppliers like NCR, who are also getting in on the act with dongles and apps under the brand of Silver, as well as a whole raft of lookalikes including iZettle, mPowa and more, but the business models are different.

iZettle has struggled to get the right terminal out there, as Chip & PIN is more challenging and costly than magnetic stripe payments, whilst mPowa is focusing upon empowering banks to engage merchants in their mobile payments revolution.

This is why mPowa questions the Square and iZettle models, saying that if you seek small merchants, then the volume of transactions is too low to be a sustainable business model.

It would certainly be true that if 90% of your clients only make one low value payment transaction per month, then the model is questionable.

However, I am sure that Square would refute such critique by claiming that the long tail of payments creates enough volume to be profitable.

It certainly seems that way, as there are a whole range of other companies competing in this space such as PayaTrader, Intuit with GoPayment, Bancard with PayAnywhere, Verifone, Payleven and Sumup all noteworthy.

Virtual currencies

WE HAVE SEEN MANY EXAMPLES of data as a currency arising, especially as the Internet era has taken off.

For example, over a decade ago, there were many virtual currencies emerging for the Internet age, including CyberMoola, Cybercash, Digicash and PocketPass, with the most successful being Beenz (1998-2001) and Flooz (1999-2001).

These sites enjoyed brief but flawed success.

Interestingly, the main reason for failure is attributed to the fact that none of these sites had enough availability, usability or recognition to gain critical mass and therefore widespread adoption.

Since then we have seen other virtual currencies arise, specifically the Linden Dollar as part of the gaming world of Second Life. Again, these emerging digital systems failed, often due to a lack of governance or regulation. For example, the Linden Dollar failed because the developers did not regulate the managers of the currency.

The most recent attempt to provide a good alternative, that gained significant traction, is Bitcoin.

Bitcoin is a digital currency designed to be controlled through encryption rather than a centralised authority. Operating in exactly the same way as cash, Bitcoins are fully exchangeable as an anonymous form of currency in real time across the Internet and, shortly, at Point-of-Sale.

The core features of Bitcoin are that they can be:

❏ Sent to anyone with a Bitcoin address;

❏ Accessed from anywhere with an Internet connection;

❏ Anybody can start buying, selling or accepting Bitcoins regardless of their location;

❏ Completely distributed with no bank or payment processor between users (this decentralisation is the basis for Bitcoin's security and freedom); and

❏ Transactions are free (for now – this will change).

Bitcoin is a fully encrypted, digital currency which, when you have some, can be used globally as easily as cash. It has no central issuing authority and, if you trust it and can use it, means that you can trade anywhere, anytime with anyone, with no interference.

Technically, it's an open-source payment tool. Like BitTorrent – a peer-to-peer file sharing protocol – Bitcoin allows the peer-to-peer sharing of value securely, globally. The problem is that authorities do not like open source P2P services like BitTorrent as it undermines traditional forms of commerce as the currency can be used for both good and bad things.

How does a bank make money out of being social?

In fact, most of the association of Bitcoin are with crime, according to government authorities, being used for drugs and terrorism. This is not actually the case, as Bitcoin does not fuel crime, just as the Internet does not fuel crime. Because the Internet enables links to drugs, gambling and pornography, that doesn't mean that you need to ban the Internet. In the same way, if Bitcoin allows criminals to trade in drugs, gambling and pornography, it doesn't mean that you need to ban Bitcoin.

And authorities cannot ban the Internet anyway as, like Wikileaks, Bitcoin is a decentralised P2P service that exists globally through any Bitcoin user's PC. These encrypted coins are then traded online via exchanges. Once you buy a Bitcoin, you can then use it to buy and sell goods and services online anywhere, with no central bank, tax authority, and bank or payment system involved. In other words, Bitcoins are digital tokens that can be traded anywhere by anyone without barriers.

This is why it concerns governments as most other forms of value exchange can be controlled. Visa, MasterCard and PayPal have offices in America for example, and are therefore controlled by the American authorities. Bitcoin does not sit within American, Chinese, Russian or any other soil; it just exists across the worldwide web and is therefore uncontrollable.

Nevertheless, Bitcoin still has to gain critical mass acceptance as a currency to buy and sell, and that's where its major hurdle has yet to be overcome. For example, in a recent academic paper reviewing Bitcoin[6], several key inhibitors to data currencies were identified:

❏ Improper usage;

❏ Superior alternatives;

❏ Government blockage;

❏ Abandonment due to deflation;

❏ Technology failures leading to theft or a loss of anonymity.

---

6   *Bitcoin: An Innovative Alternative Digital Currency* by Reuben Grinberg, Yale Law School, December 2011

Similarly, the European Central Bank (ECB) issued a report on Virtual Currency Schemes, with case studies on Second Life and Bitcoin.

In the report, the ECB calls Bitcoin "the most successful — and probably most controversial — virtual currency scheme to date."

The ECB goes on to say that the concept of Bitcoin stems from the Austrian school of economics, where business cycle theory developed by Mises, Hayek and Bohm-Bawerk floated the idea that virtual currencies could be the starting point for ending central bank money monopolies.

Why does the ECB bother to write a report and an in-depth analysis of Bitcoin and other virtual currencies? Because they are worried that they are unregulated value exchanges that could represent a challenge for public authorities and have a negative impact on the reputation of central banks.

To be honest, whether Bitcoin takes off or not, virtual currencies are going to explode thanks to in-app gaming on the mobile Internet, something the ECB report misses.

Then the real killer is when the major payments processors move into the virtual payments space. For example, Visa recently acquired PlaySpan, a company with a payment platform that handles transactions for digital goods; whilst American Express has purchased Sometrics, a company "that helps video game makers establish virtual currencies and … plans to build a virtual currency platform in other industries, taking advantage of its merchant relationships."

So already the world of gaming is developing our new world virtual economies into new world virtual payments processors and what all of these developments demonstrate is that the rise of digital currencies offer real value in the same way as cash and other hard currencies.

In other words, data is a currency and the Digital Bank needs to include data as a currency, virtual currencies, mobile money, proximity payments and more in their social money and payments plans.

So as we move into the melding of online and offline into

real-time, we could see a major shift away from bank infrastructures such as Visa, MasterCard and SWIFT into new real-time infrastructures such as VocaLink's Faster Payment services. Ironically, it was the UK government that forced real-time payments onto the banks through the Office of Fair Trading. VocaLink is the UK's ACH and owned, naturally, by the banks.

In summary, social money is all about enabling the exchange of value between individuals and businesses through electronic channels.

This exchange can be:

❑ Formal, through electronic money transfer systems such as PayPal or exchanges based upon backing from valuable metals such as gold and silver; or

❑ Informal through exchanging real money into other forms of value, including virtual money and complementary currencies.

All of these are Internet-fuelled and managed, to support the wider sphere of social finance, networking and media.

Social lending and saving

THE EARLIEST EXAMPLE OF A NEW SOCIAL FINANCE MODEL was launched by Zopa in the UK in 2005, creating a new form of lending. Since then, Zopa-style lending services have appeared everywhere, from Smava in Germany to ppdai in China to Prosper and Lending Club in the USA, as well as more and more UK domestic competitors, such as Ratesetter, Folk2Folk and more.

The basic premise of most social lending is an eBay-style platform for savers and borrowers. The savers' savings fund the borrowers' borrowings, just like a bank, but the difference is that these businesses offer no financial management themselves. They are just a platform to connect savers who want a higher return on their savings, and borrowers who want a lower interest rate on their loans. These platforms just connect them and claim that there is no financial management taking place.

Zopa, Smava, ppdai and related firms are therefore like eBay. eBay does not offer a product but merely connects buyers and sellers. In the same way, these firms purely connect savers and borrowers. This is why Zopa falls outside the remit of the regulatory authorities, although this is changing as from 2014 when formal industry regulations will be introduced.

Social lenders also win out because they are small and nimble, with few staff or overheads, which means that they can offer a service with minimal spread on the difference between borrowing and saving rates. As a result, they undercut bank rates significantly with a 0.5% spread.

This is why social lending works

IN TERMS OF MANAGING THE RISK on loans, most of this risk is managed by using traditional credit vetting services of banks, such as Equifax and Experian, and is the reason why the service has a minimal default ratio averaging 0.7%. The ratio lowers to 0.1% for the lowest risk lending operations, which is actually a better risk model than traditional banks achieve.

Typical lenders ages range from 20 to 88 years old, with an average age of 40. The amounts loaned range from £10 to £100,000, and the incentive for lenders and borrowers is to receive better interest rates than they would receive from traditional high street financial providers.

In the UK, Zopa and the other lending companies have been doing particularly well thanks to the credit crisis. For example, in 2013, Zopa's loan volumes were running at £20 million a month compared to just £6 million per month in 2010. The company anticipates processing over £400 million of loans in 2014, more than double the amount for 2013. This increase in business is also attributed to the issues of September 2008, as the credit crisis hit and UK banks HBOS and Lloyds merged, whilst the other major bank, the Royal Bank of Scotland, had to be bailed out. This meant that the traditional sources of credit access shrank considerably post the credit crisis, and savings rates were also perfunctory as central banks followed a zero interest policy.

This UK success has therefore been replicated elsewhere, with firms such as the Lending Club in the USA taking over $10 million a month through their platform.

Nevertheless, this has not succeeded in all countries, with Zopa's overseas ventures not working out for various reasons, most of them regulatory; Prosper had also failed in the USA, as the SEC shut them down when they were generating around $10 million a month and they've struggled to average more than $2 million a month ever since; there are no such systems in Africa; and the only other European success has been Smava in Germany, but they needed a banking licence from BaFIN before they could ramp up their business.

In other words, the financial protection rules of most countries has made it incredibly difficult to create successful social financing operations, even though the service they offer is not a financial one. Note again that Zopa does no lending itself, it just enables those who want to lend to connect with those who want to borrow at better than high street banking rates.

Another point is that they do no marketing, but are trusted. This trust is generated by their community. For example, most of the social lending services have an open platform for discussions that are regulated by their customers. If one customer comes out and moans about bad services, there will normally be 10 others who will shoot them down if they are wrong. Amazingly, some customers spend up to eight hours per day responding to community posts. They are these websites' biggest advocates and, as such, are unpaid marketing champions for the cause. As a result, the social community support the social offer and having your customers as your community moderators not only means less staff dedicated to monitoring the conversation but also faster response times and more trust in the answers.

Twitter – ideal for customer complaints?

ALL IN ALL, this is why these firms use social media for customer servicing extensively, and state that Twitter is one of the best service tools they've ever seen. Why? Because complaints are short, fewer than 140 characters and can be quickly handled. Pascal would be pleased as he said I have made this letter longer than usual, because I lack the time to make it short (Je n'ai fait celle-ci plus longue parceque je n'ai pas eu le loisir de la faire plus courte).

Not just that, but their staff think it's cool to be using the latest technologies to deal with customers, not some boring old call centre operation. Furthermore, it's in the public domain and that sends a powerful signal that they're not trying to hide anything.

The social tools have built a trust that since the crisis has enabled them to compete effectively against traditional banks. In fact, traditional banks weren't particularly liked before the crisis, but they were trusted because they were old, big, safe, secure and reliable. Now, that's not the case. Their image has been tarnished and the social lenders and financial firms have worked really hard to ensure they can compete. For example, most have a completely transparent business model aligned with their social community platform. They position themselves as being 'available to talk' and pride themselves on being responsive and attentive.

But this has to be a complete commitment as you cannot be public some of the time, and private some of the time. Because it's in the public domain, you cannot follow that line and then ask the customer to take something offline.

Suffice to say there is a world of direct social lending out there and it's now big business (Zopa recently gained over 2% market share for all UK personal retail credit lending).

It is not without issues, however, with the US government closely monitoring the way these sites operate and, in some cases, shutting them down. The problem with social lending services, as cited by the SEC, is that loans are not being repaid and this is the challenge for these sites: to get enough liquidity to be able to cover all the borrowings required, and to manage the risk of those borrowings.

Many of the social lending sites cover these risks by using Experian, Equifax and other credit rating agencies, or through the offer of insurances, but any squeeze on funding strains the social lending business model.

An example is called iWish from ICICI Bank. iWish provides a means for users to create a savings goal, such as saving for a car or a college education or to pay off a mortgage. As they save towards their goals, they can tell the world what they are saving for on their social network page, blog or website. As a result, their friends and family can contribute towards their savings goals and this is very friend and family oriented.

The leading company in this space is the American innovator SmartyPig, which created one of the first social saving products in 2007. They now offer a direct online smart piggy bank in America, as well as partnerships with banks like BBVA for the Compass app in the USA, ICICI Bank for the iWish service in India and ANZ in Asia. In these instances the overseas banks offer the SmartyPig service under their own brands.

Social funding and investing

SOCIAL FUNDING AND INVESTING falls into a number of sub-categories, but the major areas are crowdfunding and social trading. Crowdfunding differs from social lending in that it is investing for returns in new business start-ups and, like crowdsourcing, it pools the money of the masses into a nice venture fund to get things started.

Much of the focus of crowdfunding has been around Kickstarter, the American leader in this space.

Kickstarter provides a platform for funding by pre-selling your idea, rather than providing equity in the business.

So, for example, you have a conceptual music idea, and you pre-sell the idea through Kickstarter with the hope of getting enough monies to fund the implementation of the idea (unlike other sites where you get an equity stake in the business).

Kickstarter kicked off in business in April 2009 and, three years later, had seeded $200 million in funds across 50,000 projects.

Most of these projects are related to entertainment and the arts (about 60% of all projects), although some are technology and related fields. For example, in their most recent success, Kickstarter generated $10 million in funding for a new venture called Pebble. Pebble is a smartwatch that will connect to a smartphone. According to the Wall Street Journal, it raised more than $1 million in its first day on Kickstarter (April 17 2012) based upon an offer to pledge $115 to pre-order the watch.

By mid-May 2012, Pebble had achieved its goal of raising $10.27 million. The funds were gained from 68,929 people, making it the most crowd funded start-up ever in dollar terms at that time.

There is nothing like a cool gadget to get people excited.

Nevertheless, as mentioned, most of the projects are related to music, film, art, theatre, design and publishing, and these provide some interesting stats.

For example, of the platform's 7,388 successful music projects in June 2012, 6,446 of them (87.3%) had raised $10,000 or less; 238 of them raised more than $20,000; eight raised more than $100,000; and one (Amanda Palmer) raised more than $1 million (stats from Billboard).

Similarly, not all projects succeed in raising the funding required. In fact, about 41% of the projects listed fail with only half getting the funding needed.

However, crowdfunding is big business – Kickstarter had $100 million of pledges by May 2011 increasing to $250 million by May

2012 – and the service expanded into other regions from 2012 onwards.

That's an important development, but it's not a clear space. There are other crowdfunding services like Indiegogo, Crowdrise, Rrazoo and more, and even specialist sites for vertical markets like Med-Startr for start-ups in medicines.

And the UK already has some crowdfunding sites – like Funding Circle, CrowdCube, Seedrs and ThinCats – so it's a hot market space to watch right now.

According to Massolution, a research firm specialising in crowd-sourcing and crowdfunding solutions, crowdfunding platforms raised almost $1.5 billion worldwide in 2011, with a growth rate of 63% CAGR. They forecast that the funding figures will have doubled in 2012 to near $3 billion raised from 530 platforms, up from 452 in 2011.

So this is already a serious alternative to bank credit for business and small start-ups.

Meanwhile, social investing is illustrated by two market leading developments: eToro and Stocktwits.

eToro launched in 2007 as a social network for foreign exchange trading, and is now the largest social investment network world-wide. Over the years, it has expanded into full commodities, shares and currency trading, offering the ability to trade direct online as individuals or to follow 'star investors' by copying their social profiles and investment portfolios.

Based in Israel, the system uses real-time features to let users follow and trade based on other users' activities, and has seen strong growth, with 2.85 million accounts operating in June 2013, compared with 1.75 million in 2011.

The core of eToro is the ability to follow influential traders using a charting tool that shows every trader what all the other users are doing on the network at any point of time.

Then eToro takes that one step further so that, like Twitter, you can follow other traders on the network, watch what they do, and copy their trading activities, if you want.

The traders who get followed, as a result, have a double revenue stream: one from their trades (if they're good) and one from those who copy their trades with top traders, known as gurus, commanding as many as 100,000 followers and 10,000 copiers.

According to eToro, gurus earn as much as $10,000 a month in commissions, by attracting people to duplicate their trades, and are motivated as much by their social peer status and influence as by the money.

eToro, like other investing sites, takes a commission on the trades made but otherwise does not charge for its service and therefore offers a compelling model, as does ZuluTrade, Tradency, Stocktwits and similar social investment communities.

What does this mean for the future?

IT MEANS THAT BANKS will be componentised.

The deconstruction of banks into component pieces is already happening and is being generated faster and faster, quicker and quicker, every day through apps and mobile Internet.

It is similar to the car servicing industry.

In the 1970s, all cars were serviced by their manufacturer, who made fat margins from replacement parts and labour. The customer was forced to drive their car to the manufacturer, at the time of the manufacturer's choosing and would be forced to work to the manufacturer's agenda.

Then a firm called Kwik-fit came into the UK markets, with the view that certain car parts – tyres and exhausts – were pretty much bulk standard and could be fitted without any really specialist knowledge or expertise. Result: cars could get these parts replaced at a fraction of the cost of the manufacturer, and while the customer waited.

The illustration shows how the industry was re-engineered from the supplier in control to the customer in control, and this is what social financial firms are trying to achieve – to take the parts of banking where margins are high for low value-add, and make them accessible to all at low margins through high value-add based upon direct connections enabled by today's technologies.

This attitude is driving into all aspects of banking from payments to loans, cards to deposit accounts, and treasury to derivatives.

So the banking industry is being componentised and commoditised through apps and new social business models and structures.

Chapter 8: Designing digital banks for humans

THE DEBATE STARTED in the 1990s when banks believed they could close branches, reduce back office processing and move people and companies towards Internet banking.

It was sound logic, but was too much too soon. People and companies were not ready to commit to pure Internet banking, bearing in mind we were still using dial-up lines in those days, but that changed as the world changed, with over half of Americans using Internet banking and a third using mobile banking[7] today.

Many bankers believe that branches are the foundation and payments the life blood of the bank. The branch being critical because it provides a physical point of interaction.

That physicality acts as a security blanket as, when push comes to shove, you want a place to go and see someone and talk and also to make sure your money is there. A branch provides a place for service, and that is critical when you need that service. For example, when a family suffers bereavement, the branch is a key support mechanism to sort out the financial affairs of the deceased.

And it's also a case of choice. Some people may not want to visit branches, but they want them to be available. For example, 88%

---

7   *51% of US adults bank online*, Pew Research Report, August 2013

of customers are more likely to choose a bank with multichannel capabilities, including branch.

Moreover, the real reason why branches are required is because customers want to engage in a human dialogue.

You could do business via video, but most do not want that. They want to sit and talk and network and engage with a human, face-to-face.

This is why they want branches.

In particular, they want branches because dealing with money is frightening. It's not an easy thing. It's scary. People need help with managing their money and, for them, the branch is the place to go.

Not everyone will do this online and remotely these days, as they want to have someone to talk to about money, and that's what the branch gives them.

It is the reason why Virgin acquired Northern Rock's branches and why Marks & Spencer is opening branches with HSBC and why Tesco is opening branches.

Without branches you cannot grow a banking business, and you wouldn't pay millions for bricks and mortar branches if the bricks and mortar branches did not matter.

This is all well and good, but those who are anti-branch say that they were designed in the 18th century for a market of 300 years ago and are not fit for purpose today, as the world has become digitised and the next generation customer thinks very differently.

We are producing more knowledge, information, advice and support online than ever before.

As Google CEO Eric Schmidt said (2010[8]): "There were 5 exabytes of information created between the dawn of civilization to 2003, now that much information is now created every two days, and the pace is increasing."

This is why all the statistics show that the number of transactions

---

8   Techonomy conference, Lake Tahoe USA, August 2010

and value of services in branches are falling rapidly, being displaced by online and telephone services. Furthermore, a basic metric of banking is that a branch is a high cost overhead that is city-focused.

The city customer is in fact subsidising the rural customer, by supporting low traffic branches in more suburban areas to be provided at the cost of the high traffic branch in downtown High Street.

We are also in an industry that is in terminal decline. For example, most banks are afraid of meeting customers in branch, because there is no audit trail of what happens. That is why we can fall into the trap of miss-selling, vis-à-vis the UK's £20 billion-plus Payments Protection Insurance (PPI) scandal, especially when the person in a bank branch is a low-paid teller and not a highly informed advisor. These exposures can only grow when face-to-face interactions take place with no record. This is why remote digital service is so much better, as you can keep a full audit trail of the telephone calls and clickstream of the customer. Add to this that you can delight a customer through a remote digital experience, and that is the reason why you find the advocacy of remote banks, like USAA and First Direct, being far higher than branch-based banks.

In fact, the only reason digital customers go to branches is because the bank needs them to visit for regulatory purposes. Know Your Client (KYC) rules state that customers have to physically go to the bank to show their passport and utility bills for an account opening.

The truth is therefore, that in today's world of digitisation, people only visit a branch because the rules make it that way. If you stripped away a lot of the rules, few would visit a branch frequently. In fact most would avoid the branch, given the chance.

Designing digital banks for humans

Building a customer advisory bank

So HERE IS THE CHALLENGE for a digital bank: how to on-board clients with proof of identity, a passport, when you don't have branches.

The obvious view might be to have the customer scan and email a picture, or send a picture of their passport from their phone but, for most banks, in most economies, that is not acceptable by law. The bank has to physically see the passport for KYC and money laundering rules.

This is at the heart of the debate about why banks need branches, but there is another solution.

If it is inconvenient for the customer to physically come to see you, why don't you go and physically see the customer?

This is another alternative for the branch closure discussion, where we will see banks switch from a branch structure to a direct customer advisor operation.

In this new digital world, customers handle all of their transactional needs via direct, remote self-service. Cash is all automated via ATM infrastructure and the only residual service is sales and advice.

If the only residual need for face-to-face human interaction is for sales and advice, then why not take the best branch advisory people and give them a car, an iPad, an iPhone or Android and send them out on the road?

Whenever a customer clicks or touches 'make an appointment' on their online or mobile banking service, they can immediately see a schedule of available times from the direct advisor group and, like booking an appointment with the Genius Bar in the Apple store, you choose your time and location.

Automated online appointment systems have been around for a long time, although many banks still have not deployed such systems.

This future view would mean that 90% of all bank operations are managed by the customer and, where they need help, they can still deal with humans either remotely via telephone or directly via an advisory visit to their home or office or, if more convenient, by visiting the banks uber-branch in the nearest city centre.

By eradicating most of the branch overheads, a bank could offer this to the mass market and make two strategic gains overnight: cost reduction with increased customer centricity.

Social and banking

BANKS VIEW SOCIAL ACTIVITIES as unusual, different and new, and perfect for PR and marketing purposes.

This is a mistake, as social media can be used as a customer advisory channel.

This is illustrated well by ICICI Bank in India, which uses Facebook for full access to online banking services as well as a customer service capability.

Sujit Ganguli, Head of Brand and Corporate Communications at ICICI Bank, accepted that "our customers were worried about going into their bank services from Facebook but, as we made clear to them, you leave Facebook and are using our usual secure servers once you accept and install our Facebook app."

In other words, Facebook apps are like mobile apps for banking: it's just another app-based banking service.

ICICI Bank is using Facebook to engage their Indian customers.

In ICICI's case, they offer gaming and more to engage the customer and launched their Facebook service in February 2012, achieving a million likes in just over 10 months and over 2 million after just 15 months (there are 65 million Facebook users in India).

The Facebook page offers full service banking as well as lots of other services of customer engagement but the neatest and nicest part of the use of Facebook banking at ICICI Bank is the impact of social media engagement with the banks customer base.

Before using Facebook for customer engagement, 24% of the online mentions of ICICI were negative and only 19% positive. Now,

49% are positive and just 6% negative. That's a massive change and shows that a key part of the use of Facebook banking is for customer servicing, not just engagement.

A bank that truly engages with advising customers via social media is different today but will be the norm tomorrow.

Banks will become part of the social community, not just using these capabilities as PR or marketing mechanisms.

And that really is the difference.

If you really want to see how social media can move from social marketing to social advisor, then banks that leverage these remote social channels for social relationships will be the ones that really excel.

Do customers want branches?

THANKS TO THE RAPID TRANSITION towards continuous digital, remote access, branches are clearly becoming less important today and customers are voting with their feet by not visiting branches.

This was clearly demonstrated in a presentation given by Eric Mackor, Head of Channel Development at the Netherlands retail bank ABN AMRO at an EFMA conference in Brussels in June 2013[9], the changing dynamics of channel usage within the bank clearly showed that mobile is now their dominant contact point with customers. ABN AMRO Netherlands now receives over a million interactions per day with customers via mobile, representing three out of every five customer contacts and far outweighing Internet and call centre services.

This statistic is backed up by others. For example:

❑ Barclays Bank took 13 years to achieve two million customers using Internet banking but took just two months to reach that number for mobile banking;

---

9   "Proof that mobile banking is killing the branch", *The Financier*, June 2013

❑ Société Générale took 10 years to achieve 20 million contacts per month through Internet banking, but took just 18 months to achieve that number through mobile bank services.

With this rapid transition from bricks and mortar contact to remote, digital contact, it is not surprising that branch numbers are declining, although this varies by region.

For example, over half of all the bank branches in the Netherlands have closed in the past decade.

You find similar numbers in many other developed economies. For example, Danske Bank noted in a presentation in 2011[10], that teller transactions declined 32% between 2009 and 2011 as mobile Internet banking increased. This resulted in over one in 10 branches (11%) closing during that period compared with three out of 10 branches in Iceland. Even more notable, mobile Internet banking is impacting call centres, with calls to the contact centre in Danske falling 14% during this period. This is because balance checks and payment transfers can be achieved easily on a mobile device with an app: who needs to make a call?

All of this is reconstructing the core structure of the bank, its channels and products. For example, ABN AMRO Netherlands state that, on a weekday, a small branch welcomes 20 customers per day. Each of these small branches has an average of two staff servicing 20 customers in an eight-hour day. That's just over one customer served per staff member per hour. That's either amazingly personal service or an awful lot of thumb twiddling during the day waiting for something to happen.

The punch line, however, is that although the bank can see more and more ways to service the customer out of branch, the customer's won't let the branch go.

For example, ABN AMRO organise their customers into five distinct categories:

1. Intensive channel users who use all channels regularly;

---

10   "Banks have bigger development shops than Microsoft", *The Financier*, September 2011

2. Personal contact seeking customers, who want advice and face-to-face service;
3. Self-directed people who think they can do everything themselves;
4. Passive hybrid customers who only talk to the bank when they have to; and
5. Inactive channel users who never talk to the bank via any channel.

The last two categories represent the least profitable and smallest sector for the bank, while the first two are the targeted 'mass affluent', and the mid-category is the majority of the client base.

Amazingly, the first three categories all answered the question: "I prefer to discuss more serious banking issues in person (89%) and in case of problems, I want to be able to go into a branch and speak to someone (94%)".

This is why, according to research by Forrester in 2012[11], 80% of all current accounts are opened in branches, 75% of GenY customers conclude their product purchases in a branch, and 67% of all product sales are made in branch.

This is regardless of the fact that many customers are visiting branches less, and bear in mind this varies by country. For example, only 7% of banking customers in the Netherlands visit a bank branch once a month, down from 9% in 2011 while, in Spain, 49% of banking customers still visit a branch once a month.

People get technology today not because of its gadgets, but because it is connecting their lives to the lives of countless friends and strangers.

This is why Facebook can go from nothing to a place with the population of the United States in just a few years, and why Twitter can go from off-the-radar to on-everyone's-radar in just a few months.

It is because these media help people to manage, share and organise their lives and loves.

---

11   Is there a future for bank branches? _Forrester_, October 2012

And that's what banks have to do if they are to reconnect. They must connect people to their money and finances in a simple and easy way, especially if they are to appeal to the new generation of bank consumer.

Today's new consumers see the computer and its operating system as a history lesson.

They don't care how technology works, just as I don't care how electricity works.

Mainstream media fought this technology battle ... and lost.

That's why television and newspapers are shutting down as today's media is created on YouTube and Tumblr.

Branches are alive and kicking

BRANCHES AS RETAIL STORES are definitely not dead and are unlikely to be for some time. It's just the concept of branch-based banking per se that is dead.

Banks will still need some branches for sales and relationships, and it is for this reason that the current branch system is dead.

This is very much in keeping with the realignment of the other industries.

For example, book, record and travels shops have been closing faster than a racing rocket in recent years because these stores needed to realign for the new Internet age. Customers had gone online and were self-serving, so the traditional retail store needed to do something different.

That's why most book and media stores are now coffee lounges that encourage reading and entertainment.

How would you build the branch and the branch network if you were starting all over again? First, you would probably look around and ask: "Who's got the best retail network?"

And your answer might be Marks & Spencer, John Lewis, Tesco or Wal★Mart.

Large retailing stores in hypermarkets and shopping malls.

Designing digital banks for humans

So you would select the largest shopping places and locate your main stores there.

These would be the biggest shopping malls, and the towns and cities with the largest populations.

Then you would ask: "Who has the coolest shops?"

And your answer would probably be Apple.

So you would design your mega branch stores to be cool Apple-style sales and advisory centres in the largest locations for shoppers and workers.

Would Apple build a store in every city, town, suburb and main street?

Now do not be silly.

So why have banks done just that and, even worse, still maintain that structure?

It's history, a legacy from the past. One ex-Chairman of HSBC started his career in the Kelso branch in Scotland. When HSBC was thinking of buying a Scottish bank, the joke was 'he wants the Kelso branch back'.

And it's a lesson banks are learning. The lesson is that this is broken, just like the old record store and book shop distribution system is dead.

The result is that most banks will eventually rationalise down to just one store for every 250,000 people – or one store for every large town, city and shopping mall – rather than the current structure where this is about one store for every 20,000 people.

The question then is: what do you do with the 50% of stores that are no longer needed?

Banks designed for humans, not money

SO NOW WE WANT TO DESIGN A BANK like Apple and, in talking to a design agency that created and designed the Apple stores, they talk about designing branches for people, not money.

The focus upon humans rather than money is illustrated by the

Apple store's inclusion of the Genius Bar, a children's play area and other pieces that critics thought were a waste of time.

When the first Apple store designs were announced, Bloomberg reported: "(Steve) Jobs thinks he can do a better job than experienced retailers. Problem is, the numbers don't add up. I give them two years before they're turning out the lights on a very painful and expensive mistake."[12]

However, after opening, it was obvious that the stores were engaging customers in an even more immersive, brand-building experience. Eight years later, Apple's New York store became the highest grossing retailer on Fifth Avenue.

In other words, retailing has moved from selling products or services in stores, to using the stores as a method of building a sense of community around the brand. A sense of belonging. A sense of ownership. A sense of loyalty.

These are all things that banks aspire to, and some think they have, but it is far beyond the traditional retail experience. It's about being part of an immersion in the community.

This is the answer to the question: what are branches for?

Branches are banks' retail stores to engage their community.

The issue today is that branches were not designed for this purpose. They were designed to look after money and process monetary transactions. They were designed to handle physical forms of cash and cheques, as secure transaction centres.

This is the core challenge of why everyone thinks branches will disappear.

Because they are not retail stores engaging the brand community but transaction centres run like some administration process.

So let's start to imagine how the branch experience becomes a retail experience fit for 2013 and beyond.

The radical news about this design is that it removed the teller counters – the traditional barrier between customer and agent

---

12   "Sorry, Steve: Here's Why Apple Stores Won't Work", *Bloomberg*,

– and started to open the dialogue towards a more human, face-to-face conversation.

This new engagement is allowed because the cash disappears, as does the security issue, and far more of the branch front-office focus is allowed to be about the customer dialogue and interaction.

It then struck me that banks like Caja Navarro and ING Direct were instigating community engagement by having open house sessions. Caja Navarro in Spain offered evening classes in their stores (branches) including hair styling and flower arranging, whilst ING Direct in the USA was offering sessions where anyone could just ask questions like: "how does a mortgage work?"

Umpqua Bank in Oregon did something similar, where the bank could be booked in the evening for cocktail parties or business meetings, rather than leaving the branch space dead in those hours.

This is because there is no money in the branch anymore; the money is in the data.

Then we get to thinking about the remote experience, and the fact that most techno geeks believe no-one wants to visit a retail store, branch, anymore because they can do everything remotely.

This may be true but, when we talk about designing banks for humans rather than money, we also want to design banks that engage with humans, not distancing them.

So the bank that moves to remote servicing must still find ways to get the customer to feel a relationship, and you don't feel or have a relationship with your mobile or tablet computer.

You have a relationship with the apps on your mobile and tablet, and with the people you socialise with through those apps.

You have a relationship with the friends you call, and some of those friends might even work for your bank.

This is the secret sauce of remote banking leaders like First Direct and USAA.

They designed their operations specifically on the basis of being remote, but want you to feel a relationship with their brands by the amazing service you receive.

Their remote services are designed to be as simple, easy and convenient as possible, and when you need to talk to someone, then that someone is local.

They are customer focused and engaged, and deal with you as a human and not an account number.

They work without scripts and think on their feet, and on their handset.

That is why these financial providers consistently get stratospheric customer service results.

So let's now combine the two worlds: the retail store and the remote experience.

That is what Apple has successfully achieved through stores and online services such as iTunes and the App Store with its accessories.

If you ever get confused, you can just go ask a Genius how it works in the App Store on the main street.

So the bank designed for humans will not have retail stores that are geared for transactions, but will have retail stores that reinforce a sense of belonging to their brand community online.

Their brand community will be the community of people who are fans of their apps and services on mobiles and tablets and laptops.

They may be fans who use the brand in augmented services, like Google Glass, to see if they can afford things as they cook, commute, shop, search, work and exercise.

These fans see their financial service embedded in their daily lifestyle, not as something that is transacting but as something that is advising them at their point of living.

And every now and again, they feel prompted to go and ask: how does this work or what do I do when and it gets them into a human contact at the bank's Genius Bar in store or on the telephone.

It's designing banks for humans, not money or data.

That's the secret sauce for today's 21st century bank.

But the core of this design is using up-to-date data to service humans.

Building human relationships through digital channels

One of the big things about the Digital Bank is that it must be a human bank. Just because the bank is digital does not mean that it is automated and robotic. In fact, it means the opposite. A Digital Bank, in fact, has to be more human than a branch-based bank, as it needs to exude intuitive and intimate customer understanding through technology. In order to do this it must really understand the customer and the human interaction with technology. This is the secret sauce that makes the Digital Banks of the 21st century far more competitive in their execution of technology than the banks that simply add the technology as an adjunct to their branch operations.

Specifically, Digital Banks need to advise customers about finance by providing strong customer engagement through remote channels and so, rather than KYC – Know Your Customer – Digital Banks need to focus upon KYCC – Know Your Customer's Context. This is the leverage of location-based services combined with data leverage of the customer's needs, wishes and challenges.

The Digital Bank therefore engages with the customer through the customer's preferred access. This may mean that the bank offers signing using Facebook, Twitter or LinkedIn, where an account opening takes place in under a day using the KYC from the customer's existing bank. Bear in mind customer interfaces have changed from keyboard to touch. This also means rethinking the bank offer, moving from click-to-see to scroll-to-see. This is a key point in the tablet age: touch screen users do not click. They scroll and slide.

Mobile tablet computing has made Digital Banking far more focused. For example, placing a clear contact telephone number on all web and mobile pages is critical, as you want to encourage the customer to call and have a conversation, not just be left to scroll, slide, touch and type. After all, if you want to encourage customer interaction, what better way than making sure they can call you from their mobile screen than by providing a one-touch to call?

These nuances are all part of the theme of designing the Digital Bank from a human perspective, and designing the bank for the way in which the customer wishes to interact.

In fact, a good way to contextualise this is to view how mobile adoption has taken place in five waves:

❑ The first wave was when people used mobiles to find information;

❑ The second wave used mobiles to transact;

❑ The third wave allowed interaction with the financial provider through remote services;

❑ The fourth wave, where we are today, allows anyone with a mobile smartphone to manage everything on the move, 24/7/365; and

❑ The fifth wave arrives in the very near future where everything communicates in the Internet of things to find you.

This is why we do not have channels today. There's no such thing as multichannel, let alone omnichannel: we just have a reality augmented by digital services.

This is why the Digital Bank has to go further than just allowing customers to design their financial management as they see fit; they have to apply different views of account management by account segment.

The customer designs the bank for them, rather than bank designing for one size fits all.

It means the bank has componentised so that the customer can allocate and integrate components of the bank in a way that suits them.

But that's a long way off, as most banks today have transactional-style systems presenting transactional-style statements.

Many banks have yet to move beyond a credit and debit view of a personal financial management to a mobile financial management view.

Designing digital banks for humans

Only a very small number of leading-edge banks are offering the true view: a view that suits the way in which each and every customer uniquely wants to see their money.

And the core of everything comes down to human design, which is why we have human interaction at the heart of everything. Human interaction is the reason why 90% of account openings come through branches whilst 90% of account services are self-served.

And if we can design more of the humanity into those remote transaction services, then eventually we may just be able to design customer experiences where 90% of account openings also come through remote services.

Human touch

WHEN DESIGNING OUR DIGITAL BANK with a human touch, the processes need to be designed by true designers who understand anthropology, empathy, engagement and the human experience of technology.

The objective for the Digital Bank is to be open to the customer 24/7. Customer needs are not 9 to 5. They have financial needs, wants, desires and thoughts and want to meet them at their convenience.

This is all about deep data drilling and the reason why banks are engaged in data wars, with the data being the key. Nevertheless, deep data drilling to proactively, predictively serve the customer could be creepy. That is why the start point of the humanised Digital Bank is to ensure that this is all performed on a permissions basis.

The permission is upfront and easily accessible at any time, and tells the customer that you are using their data to give them the best service possible. This means analysing the data, leveraging the data and potentially sharing the data with other division of the bank, partners and service providers.

In return, this means that the customer will get better service with offers and discounts. It is a value exchange where the customer shares their data to get more value from the bank in return and, as mentioned, the customer can easily opt-out at any time.

The key is getting the permission of the customer to leverage their data to give them better service.

Once the bank has the customer's permission to leverage their data, then the core message is about designing for humans rather than for money.

Digitisation of the relationship needs harmonisation with the real world.

We do have real world needs for advice and support and always will.

We just need to marry those real world needs with the fact that our digital footprint today can augment and enhance our relationship with money, and with our bank, far more than it ever did before.

And, for a bank, the biggest problem is how to harmonise the data and digital analysis across their legacy which was built for channel silos, rather than multichannel integration.

Chapter 9: The implications for banking infrastructures

ONE OF THE KEY THINGS that crops up regularly is the old multi-channel nugget.

Regular research shows that banks are seeking to harmonise and improve cooperation and consistency between branch, Internet and call centre channels as their top strategic priority.

Part of the reason for this is that the bank wants to show one face to the customer and more importantly one account balance. Many of the channels operate on different cycles so if money is taken off one channel it maybe many hours before the new balance is reflexed across all the customers channels. To adjust to real time transactions, one bank runs 'close of day' at least 10 times a day. The foundations were a batch system, which most Core systems are, so to meet real time needs the systems have to be adjusted. This is also true for a number of banking infrastructures.

Banks that think this way think the wrong way.

Mobile is *NOT* a channel.

Internet banking is *NOT* a channel.

Nothing is a channel.

Things have changed.

Our planet is populated by people who have digital personas, tightly integrated with their physical lives.

We walk the street texting, talking, surfing and interacting. In fact 6% of the population answer the phone during sex.

We geolocate our address and are smart navigated to our destination via our maps app.

We check what's on TV whilst reading newspapers on digitised screens.

We transact with each other person-to-person, face-to-face, end-to-end.

In the very near future, every single person on the planet will be connected this way and, in the very near future, they will think differently.

In fact, we already do.

This is why we don't think: "I'm on my mobile channel and I shall switch to my Internet channel later".

We don't think: "Now I'm doing digital stuff and, shortly, I'll do some physical stuff".

My digital stuff is my physical stuff.

It's all tightly coupled into mine and everyone else's lives.

This is why we have to stop thinking of channels and multichannels.

It's a fundamental flaw in bank thinking.

There are no channels, just digital.

What we are seeing in the designs of new banks is a realisation of banking not being fit for today's digital customer unless it becomes the Digital Bank.

The Digital Bank does not believe in channels, and any mention of multichannel strategies is outlawed in the Digital Bank.

In fact, the Digital Bank wholly focuses upon customer lifestyle, how they choose to relate to and communicate with their financial provider, and how to deliver augmented digital servicing as part of the customers' day-to-day lives.

This is proactive fulfilment at the point of life, rather than reactive services at the point of interaction.

Proactive fulfilment is all about recognising customer activity through opt-in services, and enriching a person's or corporate's day-to-day financial needs as they arise.

This is what we get today from Amazon and Apple. I don't need to fill in multiple forms to order that book or song, just click to pay or confirm.

The Digital Bank needs to actively seek to direct customer behaviour through rewards. Buy your goods with this retail store, and we will automatically give you an extra 5% discount if you use our bank to pay, as they are our bank's loyalty program sharing partner. Save more than 10% of your earnings per month and we will give you a "Saver King" or "Saver Queen" badge to share with all your friends on Facebook.

In fact, the Digital Bank will be the Augmented Bank as it will recognise that anything can transact with anything person-to-person, person-to-machine or even machine-to-machine.

Back to the future with Cash Back

WE HAVE GROWN UP with various loyalty programs. The 1960's gave us the Green Shield stamps, which were given as rewards at certain UK supermarkets and, once accumulated, could be turned into gifts. Now the same happens digitally, no stamps to lick and rewards including cash back and discounts based on accumulated digital points. Today in the UK probably the biggest is Nectar points. The more you spend the more points you earn. Nectar is particular popular in the UK with 19 million users. As Sainsbury's uses Nectar it is going to be interesting to see if they apply the Nectar concept to their now wholly owned new bank Sainsbury's Bank.

Santander has taken this further by offering an account called 1-2-3. Basically Santander pays you for spending money from your bank account for the following:

Cash Back	Areas of spend	Value on £100
1%	Mortgage	£1.00
	Council Tax	
	Supermarkets	
2%	Gas	£2.00
	Electricity	
3%	Petrol	£3.00
	Mobile Telephones	

This service equally allows customers to crowd source for bargaining and negotiating discounts with companies. In other words, for the first time, a bank has 100,000 customers using a service to bargain on the level of discount with that company. For the first time a company is sitting across the table with a bank, whose customers want them to negotiate their terms and conditions. Imagine having a group of companies using one telephone provider and then approaching that telephone company to see what group discounts can be negotiated? This is what Santander has done.

Digital banks can bring financial benefits to its account holders very simply. The bank knows where the money is going. So by working with those people, being paid discounts in the form of vouchers or cash back, can be negotiated. If its cash back, then the money comes in centrally and each individual then receives the cash in their account.

The Internet of things

The Internet of things is where Internet communication – both wired and wireless – are placed into everyday objects from cars to refrigerators, keys to key rings, jewellery to watches and more. Anything that can have a chip placed inside, in fact. We will all soon be wearing and watching and being monitored by chips in everything, and the vision of the Internet of things is just that: ubiquitous connectivity with everything communicating and transacting non-stop.

The key point about the Internet of things is that it will be the next big wave of change. It may take ten years or so but, just as we were talking about Internet banking in the early 1990s, and it became the next big wave of change in the 2000s, the Internet of things is going to be our next big wave of change and opportunity. This everything, everywhere connected world where everything can trade and transact is a huge opportunity and change for banks, and the banks that change today will win.

When you can put a chip inside anything and everything, you can track, trace, communicate and trade with anything and everything: cars, phones, walls, ceilings, books, posters, glass, bricks and even babies.

This means banks today should be trying to work out exactly what form of trade and transaction they will be offering and enabling, when anything communicates with everything, everywhere.

❑ What will be the process?

❑ Who will be the providers?

❑ How will security and authentication work?

❑ When will banks start deploying products and services that leverage this capability?

We can see the opportunity this change offers today. Near Field Communication (NFC) and Radio Frequency Identification (RFID) will provide the Internet of things with the ability to transact.

When we talk about chips inside everything, so that they can wirelessly communicate, those chips in everything will be RFID chips today.

RFID can only hold a small amount of intelligence right now, so it needs something to receive the RFID information and that is NFC.

The implications for banking infrastructures

Hence, NFC will become the reader mechanism in phones and other devices for RFID in the Internet of things.

Today, you buy things by taking them to the teller; tomorrow, if you want to buy something, you just read the QR code or hold your phone over its RFID tag. In addition, in the near future, the Internet of things will be driven by the mobile Internet of things, where everything is geo-located and identified by the network.

It is this network- and chip-driven transaction system that is the point.

The network-centric view, where everything is monitored real-time via the network, is the more sophisticated, intelligent and likely future scenario but the chip-based transaction system may well play an enabling short- to medium- term role in allowing the network to track the transactions.

The Digital Bank will be pervasive and not recognise channels as it purely exists in every digital space that their digital customer lives in. So why do we talk about channels? Why do we talk about multichannel, omnichannel banks, and how to deal with channel integration?

Because of history.

Multichannel tried to address the issues of the separation and gaps between channels, and technology firms offered integrators and middleware an attempt to create a single customer view on a single platform. The concept was to enable bank consistent interaction across branch, ATM and call centre and the Internet.

Even then, we were talking about the rise of 24/7 mobile and wireless banking.

We called it Martini Banking back then, because the summer party drink branded itself with the tagline: anytime, anyplace, anywhere.

Now we are talking about everything, everywhere banking, and are having the same discussion as back then.

Today we talk about social media but, back then, we also talked about online communities and the

Back in the 1990s, multichannel was an issue as customers were starting to interact over many channels and consistency was a big challenge:

Could the branch tell the customer the same thing the customer read online yesterday?

Would the call centre know what the customer has said?

Can the ATM be used to provide advisory notes on receipts?

All of these issues needed some sort of multichannel integration.

The difference is that today we are in an even bigger dilemma, as answering emails and phone calls has escalated, and now also includes sending and responding to text alerts, blog comments, Tweets, Facebook and more.

The issue is clearly highlighted by the rise in user-generated data with the amount of global digital information created and shared exploding in the 2000s. In fact, from 2005 to 2011, the amount of information being created and shared, from documents to pictures to tweets, rose ten-fold to almost 2 zetabytes (2 trillion gigabytes), and is increasing every day (see below).

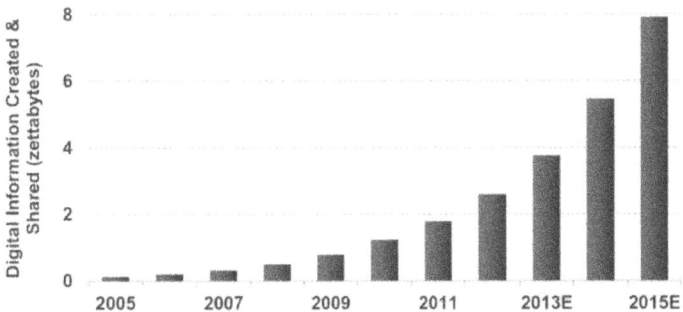

FIGURE 1: *Global Digital Information Created & Shared, 2005 – 2015E*

Equally, the demand for remote access to services via the Internet is increasing fastest in emerging, developing and growth economies (see chart on next page).

The immensity of demand over remote channels has significantly changed the mix and complexity of handling multichannel interactions, and handling them consistently is now even more important.

Rank	Country	2008–2012 Internet User Adds (MMs)	2012 Internet Users (MMs)	Y/Y Growth	Population Penetration
1	China	264	564	10%	42%
2	India	88	137	26	11
3	Indonesia	39	55	58	23
4	Iran	35	42	205	55
5	Russia	33	70	6	49
6	Nigeria	31	48	15	30
7	Phillipines	28	34	32	35
8	Brazil	27	88	6	45
9	Mexico	19	42	9	37
10	USA	18	244	3	78
11	Argentina	17	28	57	68
12	Egypt	17	30	11	38
13	Colombia	14	25	39	54
14	Turkey	13	35	17	47
15	Vietnam	12	31	7	35
	Top 15	654	1,473	15%	34%
	World	902	2,405	8%	34%

TABLE 1: 2.4B Global Internet users in 2012 — 8% Y/Y Growth, Driven by Emerging Markets. All data is from June 2012, except China which is from December 2012. There are now over 2.4 billion global Internet users, growing at 8% per annum. Presentation of Key Internet Trends 2013 by Mary Meeker and Liang Wu of Kleiner Perkins Caufield Byers, May 2013

Consistency of channels is a critical part of the Digital Bank

DIGITAL BANKS DO NOT SEPARATE their channels but view their combined digitisation as a way of providing augmented and consistent servicing. When the human elements are combined with the digitisation, it creates an augmented whole of holistic service.

This is a challenge for incumbent banks but I did find one or two examples of banks that are evolving and adapting their offers to support the augmented digital realities of today.

For example, Wells Fargo in the USA provides personalised ATM messages. Their ATMs include customised screens based on customer preferences, and favourite activities are highlighted on screen based on previous usage of the ATM by the customer. There is also a tool called *ATM Cash Tracker* which allows customers to visually track their monthly withdrawals. The tool allows customers to set a monthly withdrawal target and see details about how much they withdrew the prior month, as well as their average withdrawals per month over the past year.

This delivers a consistency across channels from personalised ATMs to personal financial management.

Now imagine what a core banking system in the digital age would do. It has all your transactions and can easily provide you with advice, each and every time you go to make a payment. The deposit account system knows how well you are doing financially. It can also help you manage your financial life. The current deposit account structure evolved from an inventory list of cheques being cashed. Each cheque in the 1970's had the cheque number on the statement. This then went to ATM transactions, direct deposits and now mobile payments. However you can't do what Wells has done with its ATMs. The ATM network is its own silo, so it can be changed to help people manage their money better. The digital age will see the core banking systems become far better at helping people and companies manage their monies better.

The implications for banking infrastructures

Another leadership in innovation can be seen at Alior Bank, Poland.

Alior Bank launched just in November 2008, with the aim to gain market share of 2–4% of the retail banking sector in Poland by 2012. By November 2012, the bank had more than achieved this aim with 4.0% market share in mortgages (new volumes), 3.2% in consumer loans and 3.8% in current accounts. It has 1.4 million clients and is the #1 bank by new client acquisition, breaking even after just 22 months. By comparison, the branch-based bank start up in Britain, Metro Bank, made losses of over £100 million in its first three years of operation.[13]

Alior Bank launched on the Warsaw Stock Exchange in in December 2012 with an initial share offer valued at 57 zlotys ($18). The bank's share offer proved to be the largest ever IPO by a private company on the Warsaw Stock Exchange and, by August 2013, the shares had almost doubled in value, trading at 100 zlotys ($32) each.

One of the keys to Alior's success is their virtual bank offer. This is a fully virtual, 24 hours a day, 7 days a week online bank branch, with a complete variety of contact options including video conversations and video chats.

The virtual branch not only offers conversation and advice, as customers can buy products and request consultants to complete application forms for them. The forms are displayed on screen as they are processed, so the customer can easily follow the progress of the application.

There are dedicated mobile applications for these features available at the Apple and Google stores, providing full accessibility and

---

13   Metro Bank's losses exceeded £100 million in May 2013, three years after its July 2010 launch. The bank made pre-tax losses of £45.7 million in 2012 but the bank's expansion plans will not be affected as it has the backing of deep pocketed shareholders including hedge fund giant Steve Cohen and the billionaire Reuben Brothers.

integration of the bank's services through smartphone devices. The mobile applications offer a number of additional features:

- ❏ look up of personal saving status by shaking the smartphone;
- ❏ quick domestic transfers (also instant transfers) via mobile;
- ❏ using social networks such as Facebook to transfer money; and
- ❏ money transfers with the use of the camera on a smartphone (by simply taking a photo of an invoice).

The bank's entertainment zone offers free music, movies, and cinema tickets. Furthermore, you can buy smartphones, vacations and other products with a special discount and 0% instalment plans.

This bank is on the leading edge of innovation and other banks, such as mBank in Poland, have transformed to compete.

What Alior and Wells Fargo show is that banks need to deliver social mobile personalisation and virtualisation to compete today. That is fine but it goes further than this. It goes further than just having consistency across channels and interactions. It needs deep data mining and intelligent data mining to compete in the age of the Digital Bank.

The multichannel myth

Banks have been adding multichannel and new capabilities to the core systems for years, and the mistake was building from history rather than revolutionising for the future. Well, this approach is easier and with a hammer everything looks like a nail.

We added ATMs because they reduced costs; we added call centres because new competitors were eating our lunch; we added the Internet because we thought we could close branches; and we're adding mobile because it's the latest fad for customer service.

We created mixichannel. Mixi meaning mixed up.

Instead of focusing upon these new channels of innovative offers for new products and services, we added these channels as cost reduction programs for self-service and branch rationalisation. As a result call centres, Internet and mobile services have been added onto existing operations, rather than engineered specifically for these services.

There are exceptions.

In the case of the call centre, First Direct was one of the first movers to make this channel work, and is the UK's leader in this area.

First Direct built their bank around a remote telephone based centre, rather than adding call centre to branch operations. Therefore, the difference is that First Direct has processes designed for remote customer reach, rather than a process designed for administering customer service when the branch is closed.

Banks reflect where they come from electronically

THE REASON MOST BANK CALL CENTRE OPERATIONS ask for name and account number, and focus upon balance and transaction statements, is because they view the branch as the key contact point.

The reason bank Internet services are dull and boring is because they are just automating statements online, rather than leveraging and using broadband-based social media, and their mobile services are often the same.

This is because the technology is being added to the bank focused around pre Internets operations, rather than using the technology to design a new bank specifically for that technology.

However, when a bank is designed around the technology, it wipes the floor of the competition.

First Direct is not only Britain's largest call centre based bank, but it's one of Britain's favourite banks. This is much to HSBC, its owners, chagrin, as the 'main bank' ratings are 50% of those of First Direct.

First Direct is a bank without branches designed for the channels of today, rather than banks with branches who added these channels onto their traditional structures.

In Japan, Jibun Bank and eBank are similar, and yet more innovative because of their use of mobile services. eBank has half of the Internet banking market in Japan, as a bank designed for the Internet. Jibun Bank has also stormed up the bank charts, as a bank designed for the mobile.

Banks have launched automations of 'the new things' at different moments of time. The key events being:

❏ Branches (pre-1970s);

❏ SWIFT to facilitate international payments (1973);

❏ International Cash Management (post 1980);

❏ Call centres (1980s);

❏ ebanking (1990s);

❏ mbanking (2000s).

Each launch therefore has a layer of legacy, which is the challenge for the traditional bank.

The pre-1970s bank is hamstrung by heritage. Even the 1990s Internet based bank is challenged by mobile, as their existence is not designed for that channel. For example, one UK bank launched its

Bringing design thinking into banks

As part of this change, banks need to bring creativity into banking and bank design.

The customer is refreshed and has new technology, but the bank is not. The bank is hamstrung by heritage technologies, and technology is not about interacting with organisations but interacting with humans as they live their lives. Many of these humans work for banks. That is why bank systems need changing.

After all, twenty years ago, banking was all about an account number. That is how we managed our customers, as numbers. In two decades, we have moved from account numbers to relationships and the focus of our systems today need to be upon building relationships with customers for life.

The implications for banking infrastructures

This needs new thinking as that renewal needs a renewal of thinking.

You cannot have systems designed for the internal organisation when the customer is renewed and when their technology has been refreshed.

You cannot have old style systems when customers have refurbished.

That is why so many banks are finding it's the core system that is the dead weight, with most banks having little change since the year 2000, when sorting out the date change refreshed many systems.

It is because banks are trying to focus upon how customers want to communicate and consume services, which is becoming predominantly through mobile and IT, and this complex interaction between humans, organisation and technology is the friction we live with today.

The world has given the customer the ability to create and share their lives through technology, and it is their creativity in design that is changing the process, the system, the structure and the interaction.

Banks should give thanks to Google, Apple and Facebook for enabling our customers to create and share anything at any time with anyone.

Anyone can create anything and, if it's popular, it is valuable. The biggest challenge then is:

❑ How to convert the business?
❑ How to persuade the bank to change?
❑ How to get the organisation to move?
❑ How to make the elephant dance?

The only way to do this is to cannibalise the organisation and create friction within the bank to renew the bank. You need freedom fighters to bring down the old and create the new. This freedom force is effectively the innovation group who are there to destroy the old bank organisation. That is why this group has to be separated from the existing organisation, as the existing organisation wants to keep the business as it is.

In other words, you have the innovators fighting the Business-as-Usual.

Business-as-Usual will fight any change, as they created what is there today. It is almost like going to the Business-as-Usual people and saying: what you built was wrong.

This is why Business-as-Usual is not interested in destroying the existing organisation. That is why you need the innovators to challenge all aspects of the Business-as-Usual, and work out how to regenerate the organisation. The core direction to the innovators is to focus upon the outcomes.

❑ What are the key outcomes for the customer?
❑ What does the customer want from the bank?
❑ How do they see the bank?
❑ How do they want to deal with the bank?

This is what Poland's mBank has determined to do and, in so doing, they cannibalised the bank.

mBank, Poland: a bank that killed its parent

mBANK IS OWNED BY BRE BANK and, in 2013, BRE bank determined to rebrand the bank entirely as mBank. mBank was a sub-brand of BRE Bank for their Internet banking and launched in 2000 to support their online activities.

BRE Bank launched in the 1980s and is now Poland's fourth largest bank.

In 2000, with the rise of the Internet, BRE Bank launched mBank, a pure-play online bank. As an Internet-only bank, it has risen rapidly to become Poland's biggest online bank and its third largest retail bank. Thanks to its success, mBank expanded into neighbouring markets in the Czech Republic and Slovakia in 2007.

In 2012, the bank realised that there were significant changes taking place in the banking markets due to the use of mobile social media, and decided to redesign the bank from scratch based upon four key tenets:

❏ Real-time marketing;

❏ Personal financial management;

❏ Mobile banking; and

❏ Social media.

The Internet bank has taken over the core bank. This involved investing $30 million for a complete redesign of the bank and the bank dropping its original brand, BRE Bank, and replacing the whole thing with the mBank brand, with a new logo and refreshed look and feel.

The transformation of the bank took fourteen months to complete.

The new service offers over 200 new functions and improvements. In addition to design, these include a smart transaction search engine which makes a search through your transaction history as easy as using Google on the Internet, with results displayed in real time.

Paying for things can be done the usual way, or you can make direct payments to friends and family via Facebook and text messages. Customers no longer need to remember or ask for account numbers, and can use phone numbers or social network connections instead.

Facebook transfers appear on the home page of the person you send the money to, and are just messages sent with cash attached. Personal finance management (PFM) is provided to allow budgets and alerts, where the system clearly explains when money was spent and what for, along with a forecast of overall spending by the end of the month.

An online customer service is available 24/7 via a video or voice connection.

It's the first bank mobile social solution that comes close to delivering the new retail banking reality.

In conclusion, the Digital Bank is already here with banks like mBank and others such as Fidor, Jibun, eBank, ICICI Bank, Deniz Bank and more delivering different version of new business models for today's 21st century customer.

All of these banks recognise that there is not a channel play here, but an augmented customer service play focused upon the customer's point of existence. It is delivering digitised financial services. These services are highly personalised at the point of relevance.

This is the core change that the Digital Bank demands of the traditional banks and most traditional banks are working to get there but are hamstrung by their heritage systems, heritage structures and heritage thinking. That has to change if they are going to be capable of competing in this all-encompassing digital world.

Chapter 10: The importance of data for the Digital Bank

DATA IS THE FUTURE competitive battleground for banks.

In fact, data is the battleground of today, but many banks have not realised this yet.

Some banks realised this years ago:

"Information about money has become almost as important as money itself."

Walter Wriston, CEO/Chair, Citibank, 1967-1984

"Banking is just bits and bytes."

John Reed, CEO/Chair, Citibank, 1984-1998

Yet most banks do not, not even today.

When they do, banks will move towards data being their most critical asset – above capital, liquidity, spending patterns that can go to the individual or company with complete accuracy. Banks, because of the need to borrow money and now the need for on-going compliance have data that the heavyweight data managers of Google, Amazon, and Facebook would crave.

In economic terms, what has value is based upon scarcity, and salt had scarcity historically which is why it was more valued than gold. Today, it has abundance, and hence is cheap.

Now data has abundance, and so is also cheap. But information is power, and so it's what you make of the data that has scarcity. In other words, turning data into knowledge is where the value lies today, just as turning rock and sea water into salt had value in ancient history.

As data challenges grow, solutions will emerge. Just as salt was a currency, it was not the salt that was important but the quality of the salt extracted by the salt miners from different seas and rocks.

Data as a currency

THIS WAS MADE CLEAR by Craig Mundie, Head of Research and Strategy at Microsoft, who said that: "Data is the greatest raw material of business, on a par with capital and labour. What we are seeing is the ability to have economies form around the data and that, to me, is the big change at a societal and even macroeconomic level."[14]

Now this is where it gets interesting, and brings data mining to the fore.

The essence of data mining is to dig deep to find relationships between diverse and fragmented pieces of data. In banking, it honed in on the single customer view, with the idea of deeply mining customer transactions to find out how to leverage sales.

Data mining days began in the 1990s with the whole focus being upon data usage for sales. Get more cross-sell, get more relationship depth, get more profit, get rid of loss-making customers ... nothing was really focused upon customer service in that era.

It's similar to the days of Business Process Re-engineering, where the push was for banks to reinvent the customer relationship from the external interaction viewpoint inwards ... instead, most banks opted for purely incremental improvements to internal processes to lower costs.

---

14   "Data, data everywhere", the *Economist Technology Supplement*, February 2010

Deep data mining can now be used to leverage electronic relationships that deliver the depth, loyalty and sales that banks were seeking a decade ago, and it can all be done today without a human hand involved.

On Apple, you download an app or iTune and suddenly get recommendations of a thousand others. Using their music network, every artist you listen to brings up five others you might like.

Google uses your IP address to send out local information and adverts that relate to your search history.

Visa is saying that they can now analyse two years' worth of customer transaction records, or 73 billion transactions amounting to 36 terabytes of data, in 13 minutes using cloud computing. This would previously have taken a month with traditional methods of internal computer processing and in 1990s forever.

What does it mean?

It means that they can work with banks to sense customer's lifestyles, needs, relationships and desires.

Banks can then target offers to those needs and desires automatically.

In other words, predictive marketing and financial advice are the battlegrounds of Big Data.

What is 'Big Data'?

THE TERM BIG DATA stems from the Second World War, when the phrase Big Science was used to describe the rapid cycle of changes that occurred in scientific disciplines during and after World War II.

The term Big Data was first used in the late 1990s and no-one can quite put their finger on how it came about. It appears to be through a combination of academic papers and META Group (now Gartner) who say that:

"Big Data are high-volume, high-velocity, and/or high-variety information assets that require new forms of processing to enable enhanced decision making, insight discovery and process optimisation."

It may be simpler to say that Big Data is just this phenomenon we have today, where every person on the planet is creating digital records such that we are drowning in data.

For example, we are now creating exabytes of data every day. An exabyte is a 1 with 18 zeroes after it – or 1,000,000,000,000,000,000 bytes – an incomprehensibly large number and illustrating well why we are drowning in data.

The challenge of this ocean of data is to work out how to sieve and select relevant information from the data for marketing and sales, service and advice.

This is the great opportunity of the new digital bank age: making knowledge out of disparate data because those who make sense of Big Data will win.

As one bank Chief Executive recently stated: "Our peers I can handle. We're in the same boat. But if Google opens a bank, with all their data and analytics – then we're in big trouble."[15] Google also has its own definition of privacy while banks are fined and censored if they fall short of regulations.

Banks as secure data vaults

TRADITIONALLY IN BANKING, the trust guaranty from banks has been in keeping your money, funds and investments safe and secure.

The bank can become a data miner for example, but not mining for financial data but data in any and all of its forms.

A bank may network businesses together to say that your business may benefit from a relationship with this business, based upon buying habits and supply chain and treasury similarities.

---

15   Technology take-off threatens banks foundations, *Euromoney*, September 2012

So a bank is no longer related to finance at all, but is totally focused upon being the best at utilising data and data analysis.
The challenge is that banks are not leveraging their data richness.
Meanwhile, the biggest issue for a bank in mining the data is the
silo structure of banking.

The traditional bank has locked itself into products and channels,
rather than customer and data, and the result has been a silo structure that is very difficult to manage unless the bank rebuilds itself
from the ground up as a digital bank.

Amazon as a traditional bank

The book division doesn't talk to the music division. The music
division won't talk with the electronics division. The retail business won't talk with the wholesale business. The wholesale business
won't talk with the cloud business. Oh, and the Kindle division
won't even talk with the book division!

None of them will share customer information with each other
and, as a result, no-one knows what customers buy from Amazon
as a group, when or with what sort of payment type.

There's no view on Amazon's share of wallet or who is cross-sell-
ing what to each customer.

They have tried to improve this over the years, but the line of
business heads for books, music, electronics, retail, wholesale, cloud
and kindle are all at each other's throats, motivated by their own
line of business results.

Should someone suggest a new organisational structure that
would allow seamless integration of all divisions, with a single platform for the company, to see a single version of all the customer
information, in a single view. The Board's management and Chair-
man probably see it as folly. The bank they would say has been
around a hundred years and has its traditions.

Ah, but Amazon is a not a traditional bank.

Of course, it has divisions – and divisions are meant to do just that, divide – but Amazon's divisions don't divide the organisation by customer owners but by logical structure of organisational delivery.

So what sort of business would divide customers across lines of business so that never the twain shall meet?

What sort of business could have customers caught up in several areas of their business, but dealt with as though they were all separate people rather than just one person?

What sort of business could not recognise that a small business customer might be one and the same as their premium accountholder?

What sort of business could ignore the fact that this person is living with that person with two teenagers in the family?

What sort of business would allow their loans business to stop their savings business from making contact with the customer, even though they work for one and the same company?

What sort of business would allow line of business owners to stake their turf for internal gain but at the customer's loss?

A bank.

Yes, the silo divisions of banks mean that they ignore their greatest competitive weapon: data, and of course this leaves them open to a more meaningful partnership offered by non-traditional competition.

Information as a competitive weapon

AMAZON STARTED OUT IN THE 1990S and was all about books, we thought.

Sure, that's where it started, but it soon moved from books to music, films and more.

The company then got really smart and started to make data mining its core art form.

Data leverage by looking at data prints – like fingerprints, the unique way in which each of us search, buy and consume – and then relating our data prints to each other to find relationships.

By doing this, Amazon built its business on finding offers that you might buy, because people like you buy it and they know this thanks to our unique data prints.

Soon, Amazon was more of a behemoth of data, moving into selling anything from white goods to televisions.

Recognising its information leadership, Amazon opened Amazon Web Services (AWS) to become the largest cloud computing firm out there.

Amazon now adds server systems to AWS every day that would have been the equivalent of the complete server architecture required to run the total retail business two years ago.

If Amazon was still just an online bookshop, it would be dead.

Which brings me back to banking and the issue we face?

The reason we worry about Amazon, Google, Facebook, Apple and others is because they all know how to use information as capital. They are also changing how we see the world and set standards that have to be met.

The future basics of banking: data leverage

THERE'S AN OLD CHART used to show how data is power:

The idea is that the more humans add context and interpretation to raw data, the more powerful you become. By human interpretation, it can be programmed too, but it's the human spin on data that creates leverage.

It is the silo structure of banks that stops them from gaining this leverage.

These silos of the bank cut the data cake into pieces and the more pieces, the less powerful the data becomes.

The separation of information means that the information is a lot less meaningful than the picture you would gain from the whole.

The Data Protection Act is often an easy way to say ah no, don't go there and try that.

Alternatively, the rules about Chinese Walls across the bank are put forward as the challenge.

Admittedly, with the Vickers and Volcker Rules coming into force about separation of bank activities, the latter may be a real challenge but it's not the real barrier.

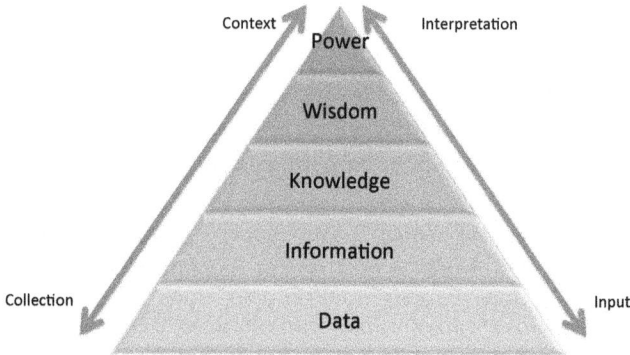

The predictive, proactive bank

WHEN GOOGLE KNOWS my searches are ideas, they can predict what is relevant for me. If I searched for headache tablets side effects, they might recommend that I switch to paracetamol, and direct me straight to my nearest Boots or Walgreens. If I happened to be pricing TVs, they might offer me a special deal as for Best Buy or PC World to get a discount.

If you don't think that Google Analytics are the key to predictive, proactive marketing, Google predicts stock market movements pretty accurately as well as flu trends, election results and more.

In the same way, banks can use transaction data combined with search trends and other data to predict and then proactively offer service in real time.

"Any sufficiently advanced technology is indistinguishable from magic."
Arthur C. Clarke, 1917–2008

The augmented economy, where everything is connected and communicating and transacting non-stop became a reality in 2013 when Google launched Google Glass.

Google glass is a pair of WiFi glasses that allow you to see enhanced information about all that is around you in real time.

This could have a profound impact on society, and therefore customers and consumers.

This is because banks do not analyse data on an enterprise basis, but usually hold this data in divisional stores organised around products, channels and lines of business, and the politics internally are the greatest blockage for change.

But without that change, banks are stuck with piecemeal data sets. These sets are being analysed in pieces, and that is not good enough today.

This is why banks need to completely architect their enterprise technologies to enable deep data mining across all data, and create semantic marketing programs that sense customers' needs proactively and pre-emptively.

Like algorithmic trading in capital markets, where algorithmic news feeds allow trading in equities to move in real-time, high frequency black box strategies that maximise returns, we're talking of applying the same technologies to retail transaction services for customer loyalty and wallet share.

Banks need to radically shake up and wake up

FRANK PARTNOY, AUTHOR OF MANY BOOKS about banking and financial markets, recently asked the question: "How many people does it take to operate a modern bank and how much should such a bank's shares be worth?"

Frank answers it by saying that "in the future, improved technology will reduce the number of human beings needed to allocate capital, as it has done in other service industries ... contrast the employment numbers for banks and technology firms."

HSBC and Google obviously differ in substance but both companies are focused on innovation and service, and also have roughly similar market capitalisations. The striking difference is that Google generates these numbers with fewer than 30,000 employees."

This begs the question: who generates more value and how? Let's take a look.

Bank of America, The Banker magazine's #1 Bank in the World 2011, and HSBC are viewed as being the best and most efficient banks in the world. What do their numbers look like when making a direct comparison with Apple and Google[16]:

- Google generates around $1.5m in revenue and $391,000 income per annum per employee;
- Apple generates over $2m revenue and $513,000 income per annum per employee; whilst
- Bank of America generates under $0,25m revenue and almost $60,000 loss per employee per year; and
- HSBC makes $0.181m revenue and just $44,000 income per employee per year.

This is the point as to why the industry has to change. The numbers just don't stack up and if there were free rein of competitive forces in banking, then the technology oriented new entrants would have decimated the banks by now.

For example, take PayPal.

PayPal's employee numbers aren't generally made available for example, but I'm guessing they employ a maximum of 3,000 people worldwide – the most quoted number is that PayPal employs 2,000 of the 17,700 total eBay workforce – and will make a revenue of $100 billion and income of around $7 billion this year.

That would equate to $50m revenue per annum per employee generating $3.5m income each.

Banks are large-scale retailers with many stores that serve consumers and businesses. Without that store network, banks would

---

16   Figures are from annual results 2011

obviously be able to run as pure digital machines, and some already are and some will be.

One of the best examples of the economics of a digital bank is eBank, in Japan, which supports over three million customers with under 200 staff.

15,000 customers per member of staff is the dream for an ultimately automated bank.

Right now, the numbers for most banks would be more like 1,500 customers or fewer per staff member.

Equally, revenues and profits per employee should be more akin to technology firms than banks although this is predicated on the basis that the banks have no branches and therefore need far fewer people. Banks have not let that paradigm shift happen yet.

It is why so many new entrants – that is if they can get a banking licence – are moving into bank' territory on the basis of fully automated offers, as the banks have left that competitive space wide open. Most people can't obtain a banking licence, the system is against them. The smarter ones use the banks and their infrastructure. PayPal is simply superb at this and in the UK Barclays does their Faster Payments under a white label agreement.

Like book retailers, music retailers, holiday retailers and more, financial retailers will start to feel the full force of the technology shift of consumers to remote banking via mobile Internet.

The importance of data for the Digital Bank

Chapter 11: The impact of data on bank processing

DURING THE PAST FIVE YEARS, almost every conference on banking and payments has talked about mobile and tablet computing.

This is because everyone thinks mobile is the hot space today, which it is, but it won't be in the near future. Very rapidly, the device focused dialogue will move on to the Internet of things as everything gets Intel inside.

This is the major disruption on the technology landscape, and here are the facts:

The volume of Internet traffic doubles every 18 months and is currently running at around two zetabytes, which is a trillion gigabytes, and most of it is video;

- ❏ The actual size of the Internet doubles every five years;
- ❏ By 2020, there will be 50 billion devices connected to the Internet, compared with 17 billion today;
- ❏ By 2020, there will be 6.6 devices per person using the Internet, compared with 2.5 today;
- ❏ Nanotech is already here, with computers and cameras at sub-1mm sizes;
- ❏ Over 100,000 telephone masts are being built every year;

❏ The number of WiFi units shipped has quadrupled in the last five years;

❏ Under IPv6, up to 52 thousand, trillion, trillion web addresses.

The Internet of things is the idea that the vast majority of our electronics will be connected to the Internet and/or other nearby devices.

A refrigerator, for example, may have a touch screen on the door and be connected to the Internet, allowing you to remotely access information such as what is in the fridge, its temperature and whether or not you have what you need to cook that spaghetti bolognaise this evening. Another example is the Nest thermostat, which is a thermostat that allows you to remotely manage your house room temperatures from an app.

In years past, this idea was just that, an idea — something we said was coming. This year, however, was the first year when we could actually see the Internet of Things on display and it is clear that, within the next five years, we will be living in a world where almost every electronic device we own will be connected to something.

What does this mean for transaction banking?

IT MEANS THAT THE WORLD WILL BE AUGMENTED, such that wherever you go and whatever you are doing, your world can be supplanted by relevant, proactive, predictive, proximity-based information.

This concept was proposed many years ago when technology firms made a play for banks to deploy data warehouses to perform predictive analytics based upon consumer propensity models of their data-based behaviours.

It may sound a little complex, but basically it was meant to use data patterns to predict what the customer would do next and whether they might need some new financial service. The supermarkets are superb at this and design grocery lists with incentives based on what you last bought.

Today, this form of marketing combined with transaction is

imperative, as predictive marketing is the battleground of Big Data, and Big Data allows payments processors to be far more proactive rather than reactive.

If you don't think that Google Analytics are the key to predictive, proactive marketing, just checkout the results of research of three academics who find that Google predicts stock market movements pretty accurately:

"Debt" was the most reliable term for predicting market ups and downs, the researchers found. By going long when "debt" searches dropped and shorting the market when "debt" searches rose, the researchers were able to increase their hypothetical portfolio by 326% (in comparison, a constant buy-and-hold strategy yielded just a 16% return)[17].

In the same way, banks can use transaction data combined with search trends and other data to predict and then proactively offer service in real time.

Your smartphone would send your preferences to the store's database as you entered such that you get special discounts on the items you buy most regularly. As you walk past these items, they change prices dynamically and your phone beeps a special discount deal today to alert you to these offers: *Just for you, buy one get one free (this offer is not available to anyone else instore).*

For example, the Square Wallet and PayPal Instore apps are good examples of the evolving shape of retail payments, where transactions no longer involve any physical exchange of tokens, cards or paper, but purely a confirmation of the payment via the mobile network.

And that means being the bank that mines data to provide predictive, proactive, proximity based payments. The bank knows who has readily available money and how much.

---

17   "Quantifying Trading Behavior in Financial Markets Using Google Trends" by Tobias Preis, of Warwick Business School, Helen Susannah Moat, of University College London, and H. Eugene Stanley, of Boston University, *Nature Publishing Group's Scientific Reports*, April 2013

The impact of data on bank processing

The net: net of all of this is that banks and all firms will soon be focused upon wireless transaction processing through the net.

In other words rather than mobile payments and mobile banking that we talk about today, we will be talking about augmented payments and augmented banking tomorrow and, in that world, money is data.

Physical cash

PHYSICAL CASH IS EXPENSIVE for the banks. Remember Mondex the initiative to provide electronic money? That died because of adoption as it was an all or nothing proposal. Mondex's best community was at a university which could create its own electronic cash infrastructure.

We have a war on physical cash because we want to replace it with electronic processing that is cheaper and easier, and electronic processing means that cash becomes data.

The problem with this view is that there is still a lot of cash around, with cash still representing over half of all payment volumes in most developed nations and increasing in usage. For example, even with all of the advertising for mobile and contactless in the UK, the *Cash and Cash Machines Report 2013* shows that cash usage increased by around 10% for payments in the UK in 2012, with the number of cash payments (by businesses and individuals) up from 20.6 billion transactions per annum to 20.8 billion, representing more than half (54%) of all payments.

But cash in the economy is going to decrease in importance over time if, for no other reason than that the processors and financial institutions are all determined to displace cash with other forms of electronic payment

In fact, the value of cash payments has been remaining steady in many economies, whilst the volume of cash transactions has been decreasing as alternative payment forms, such as contactless and mobile payments, take over.

This brings us to the core point that we all believe cash will decline over time, replaced by electronic, digitised transactions.

It means that the regularly quoted comment from gangster Willie Sutton to the question "Why do you rob banks?", "Because that's where the money is", is very last century.

Willie Sutton robbed banks physically, taking the cash out at gunpoint. Today, the gangsters take data.

It's the data that gangsters need to rob, not the money.

That's why the majority of cyberattacks target financial institutions.

According to RSA's December 2012 Online Fraud Report, 284 brands were targeted in phishing attacks during November 2012, marking a 6% decrease from October. Of the 284 brands attacked 45% endured 5 attacks or less. Banks continue to be the most targeted by phishing, experienced nearly 80% of all attack volumes.

Many people report how much like real life the phishing sites are and the wording is so customer friendly. That is because the fraudster is not concerned about the legal ramifications if the wording is wrong. Nor are they concerned about what the regulators would say if the wording breached their requirements. One bank sent a credit card to a person on death row which was promptly used to the maximum. As the debt collector later said, it is difficult to incentivise the credit card holder to pay when they are sitting on death row.

Sophos regularly report details of bank cyberattacks, and McAfee Labs researchers recently debated the leading threats for the coming year and show that it's only going to get worse:

- Mobile worms on victims' machines that buy malicious apps and steal via tap-and-pay NFC;
- Malware that blocks security updates to mobile phones;
- Mobile phone ransomware "kits" that allow criminals without programming skills to extort payments;
- Covert and persistent attacks deep within and beneath Windows;

The impact of data on bank processing

❏ Rapid development of ways to attack Windows 8 and HTML5;

❏ Large-scale attacks like Stuxnet that attempt to destroy infrastructure, rather than make money;

❏ Malware that renews a connection even after a botnet has been taken down, allowing infections to grow again;

❏ "Hacking as a Service": anonymous sellers and buyers in underground forums exchange malware kits and development services for money;

❏ Nation states and armies will be more frequent sources and victims of cyberthreats.

So when we talk about our wonderful new Internet age, the key is to realise that it's the data where the money is, not the bank, the branch or the cash machine.

Banks need to think about how they reconstruct themselves for the 21st century as new data management firms, from upstart payments processors to Internet service providers, to mobile carriers, all move towards the payments space.

The war for all of these firms is to be the best at processing transaction data as people exchange information digitally online.

For a bank, what this really comes down to is that banks are becoming pure managers of bits and bytes of data.

Most people would be at far more of a loss if they lost access to their online accounts, had their usernames and passwords changed, had their identity copied and compromised online, or similar challenges than if they lost their wallet or card.

For some, they would feel that their lives were lost if their Facebook or Twitter accounts were blocked or deleted whilst, for others, their *World Of Warcraft* gold is more valuable to them than their total real world asset base.

By the same token, the data is where we have our greatest opportunity and threat. We talk about Apple, Amazon, Google and Facebook with admiration, but the core of these companies is not music, books, search and social networking.

It's data management.

The themes all have one core point for traditional banks and payments firms, and that point of opportunity is that the bank of the 21st century is not a bank as we would recognise it at all. The bank has to become a secure data vault. The vulnerability of data, and hence the secure management of data, is where banks and processors can truly leverage their capabilities. After all, it is money we are talking about and data is more important than money, then the bank that securely manages data is the bank that will win.

This is where the radical departure takes place from last century banking.

Last century banking was predicated on money, paper and the physical transfer of goods.

21st century banking is predicated on data, context and the electronic transfer of goods and, most of all, 21st century banking is based upon data security.

The biggest fear of corporates and consumers is that transactions will not be processed properly, that their bank access details might be compromised, that their data and therefore their money may be stolen.

That is why banks have to step up to a big challenge: guaranteeing data security.

21st century banks need to be bold and guarantee customer data is secure.

The issue with this is that it would make the processor or the bank a target for hackers, but that is the exact point. Banks should beat the hackers at their own game and make bold claims, such as we guarantee your money and your data is 100% safe with us. After all, if banks or their partners don't do this, who will?

According to many in the industry, banks and even their payments processing partners are not positioned to do this. In fact, some believe that banks should leave secure data management to people who know how to do this such as Google and PayPal.

If this is the attitude of the financial community, this is where the biggest future weakness lies. If you give Google or PayPal the

The impact of data on bank processing

opportunity to become the secure financial data manager or the secure data vault of everything, then what is the role of the processor, and the bank, in that future? Surely this just gives the whole game away to someone else?

This is why the focus upon data and data security is the key to the future. It is an adjunct to the focus upon money and financial security, but data and information security that will differentiate the future winners and losers.

In the meantime, banks have to transition from the old world of physical monetary security to this new world of electronic data security. There is a transition time between the old world and the new, and the question is for how long is this transition going to take place?

By way of illustration, if you look at the new world, there are already many different models of payments and banking emerging.

On the one hand, you have visionary financial services providers – Moven, Simple, Gobank, Bluebird, Fidor, Jibun et al – pushing the envelope of being vaults for secure data. Their premise is that the leverage of data and the knowledge they can gather from your data, allows these firms to improve the value you receive from your shared electronic relationship.

There are other new models of finance emerging such as Zopa, Friendsurance, eToro and more that will change the game again. These providers are all seeking to connect people and money through social mechanisms, and base their business upon seeing new niche opportunities for managing the exchange of data value. Wonga, the payday loan company, uses an enormous amount of data to provide loans of up to $1,500 within 15 minutes, and that includes moving the money, to anyone off the Internet.

Finally, we see a few hybrid banks emerging such as Alior and mBank in Poland, which offer the mixed old and new world capabilities to reach the broadest audience with the deepest relationships. Both of these are traditional banks with branches that have rebranded and launched as hybrid banks, incorporating social data security with bank data security.

Data is the key

THE FUTURE IS ALL ABOUT PERVASIVE ELECTRONIC CONNECTIVITY. With such ubiquitous connectivity of everything from shoes to walls, doors to cars, windows to fridges, we will see non-stop capabilities for leveraging data assets for sales and servicing. The contextual capability to market and target consumers and corporates at their point of need is going to be the battleground for future transaction processor. The banks that not only leverage their data assets effectively but do this securely will be amongst the companies that win.

Chapter 12: The implications for fraudsters and launderers

THE WORLD HAS BEEN CHANGING through an accelerating digital information revolution. Inexpensive networks and technology is now available to all. The digital revolution is connecting everyone and allows them to communicate and share anything. The last revolution of this scale was the Industrial Revolution, and revolutions create revolts.

That is why they are called revolutions because they usually occur when there is progress, and progress creates riches for the few involved in the progression whilst the poor get poorer.

The Industrial Revolution created the Peabodys, Carnegies, Rothschilds, Oppenheimers and more,   whilst the Information Revolution gives us the Gates, Ellisons, Zuckerbergs and Dorseys.

Revolutions create revolts and this is as true today as it was in the Industrial Revolution.

Throughout history are many moments where the mass poor attack the elite rich to rob them of their wealth, and we see the same today.

The difference today is that data and the mobile Internet are making a big difference in allowing the oppressed or the poor to work together and break out of their oppression.

This is obvious when the oppressor is a person – Colonel Gaddafi,

President Bashar al-Assad, President Mubarak, President Ben Ali, and Saddam Hussein – all ably illustrated by the Arab Spring.

The key to this moment of revolution was technology. The Internet and mobile text messaging enabled this movement of the protest. Facebook played a particularly important role in the protest. For example, as protest rippled over into Egypt, the government began to shut down the Internet. There were only four Internet service providers in the country and they were told by security services to shut down operations. Text messaging on mobile telephones was also blocked, as the government tried to stop the people sharing information.

Then support arrived from overseas, with messages sent by Facebook and Twitter en masse offering dial-up modem links to Egyptians to continue to communicate. Therefore, even though the government had taken down all official lines of communication, unofficial technology channels sprang up rapidly to provide alternative cover, and many of these unofficial channels were being launched by individual supporters in Holland, France, America and around the rest of the world.

Is this a time when societies worldwide change their world or will media, finance and government, continue to control and direct?

Just look at Molly Katchpole, the young lady who posted a petition on Change.org to get Bank of America to reverse policy and waive the $5 per month fees they were going to impose if people used their debit cards.

The fee was to recoup losses due to the implementation of the Durbin Agreement, part of the Dodd-Frank regulatory changes in the USA. This agreement wiped out profits from interchange on debit card transactions and many US banks decided to add a fee therefore, in order to recoup losses (note: Bank of America were not the only bank to do this, just the first to get the headlines).

The move proved so unpopular that Molly's petition rapidly gained traction, was promoted by Change.org and then was picked up by major national media like the New York Times.

When the online petition reached 300,000 votes, Bank of America reversed the policy.

It resulted in the program being voted the biggest PR gaffe of 2011 by most marketing magazines and CEO Brian Moynihan admitting that it resulted in a surge of account closures.

Note the speed of this change, however, in that Katchpole posted the petition with 100 signatures 1st October and in one week this rose to 200,000.

There are many other examples of how the mobile, social Internet is impacting banks such as the Wikileaks and Anonymous attacks on PayPal, Visa and MasterCard.

At the time, American firms such as MasterCard, Visa, PayPal and Amazon, were trying to close down funding to Wikileaks, due to the websites leaks of top secret US Government information.

The leaks included films of American bombings in Iraq that killed two Reuters journalists, something the US Government had denied happening, and so the government were publicly shamed and embarrassed and used their influence on PayPal, Visa and MasterCard to stop funding for the Wikileaks services.

The real shock was the reaction of Wikileaks supporters to this action however.

Supporters of Wikileaks targeted MasterCard and brought their web services to a halt using a simple Distributed Denial of Service (DDoS). The attack targets the bandwidth of the website, sending TCP, UDP, or HTTP requests to the site until it goes down. This hit MasterCard's 3D Secure and broadband payments services, and went viral using the term Operation Payback: "an anonymous, decentralized movement which fights against censorship and copywrong."

Cyberwars: a far bigger threat

Whilst consumers are creating mobile social pressures and hacktivism, governments are creating global cyber-attacks.

The implications for fraudsters and launderers

This cyber warfare is already rife, with a host of malware targeting Middle Eastern nations.

What is obvious from these developments is that cyberattacks are the new form of warfare that evades direct hand-to-hand or nuke-to-nuke combat.

And no nation is immune from attack.

President Obama is acutely aware of cyber vulnerabilities because he got hacked himself.

We continually try to be one step ahead of hackers, hacktivists, cybercriminals and cyber threats, but we are actually always one step behind.

Like the regulatory conundrum – you can only fix the system with regulation once you've seen it go wrong – the cyber conundrum is very similar – you can only block the attack once you've realised you're under attack.

Can any company claim to be bulletproof? As security will tell you the safest computer is the one that is switched off. It is a continuous struggle of good verses bad, and the bad are often the innovators especially if it is new. Digital is new.

In recent developments in the Middle East for example, the latest systems attack bank accounts, rather than nuclear plants. This is because the banking system is the heart of the economic health of a nation, with Gauss malware targeting multiple users in select countries to steal large amounts of data, with a specific focus on banking and financial information.

So how should a bank protect itself from hacktivists and cybercrime?

THE REAL CHALLENGE for the banking system is how to protect firewalls from attack by hacktivists, goverworms and cybercriminals and, conversely, how to deliver easy access to online banking for their clients and customers.

On the one hand, everyone wants mobile access to his or her account balances and to make payments; on the other, no-one wants to consider the issue of haemorrhaging losses if they don't protect their account properly.

So there are two distinct focal points here for information security within a bank:

❑ Protecting the banks information from attack; and

❑ Allowing the bank's customers to access the information they need when they need it.

A targeted hack is a concern, and there are many instances of banks failing to deal with this properly. Last year, for example, hackers got access to some of Citibank's customer data, with at least $2.7 million lost by 3,400 customers. That's small beans and is manageable, but shows the vulnerability.

The insider threat is even greater, with employees who can gain millions by selling access to bank data. An instance of this was also seen last year, with Bank of America losing over $10 million thanks to a staffer giving away account details to an identity theft ring. Insiders, according to Gartner, are involved in 60% of bank fraud. It is that age old question, who spies on the spies.

Banks historically took fraud as a cost of doing business. Secondly if a fraud happened it was often quietly swept under the carpet as such activities could undermine people's confidence in the bank.

This was well illustrated by Sumitomo Bank, who lost almost $350 million in a keylogger scam. This is the very same bank that was fined £3.5 million by the Financial Services Authority in May 2012 for serious IT governance failings.

This means that the way in which you guard against data failings from external attack is by having the obvious data protections: firewalls, secure sign-on, dual authentication with triangulation of access, real-time business events monitoring and so on.

Banks should be moving towards much improved real-time tracking and business intelligence about their information flows, and this will alert them to any security breach. In addition the

The implications for fraudsters and launderers

fraudster, launderer and cybercriminal do not care about which bank they rip off. Banks have invested heavily in all types of anti-fraud and AML processes and people in isolation. Six UK banks have run a proof of concept of having a 3rd party review both sides of a Faster Payment transaction. The results showed:

❏ The six banks could see 70% of the network against their normal 10%;

❏ 50% of the transactions which scored highest turned into actual frauds;

❏ Nine money networks were discovered and are being investigated.

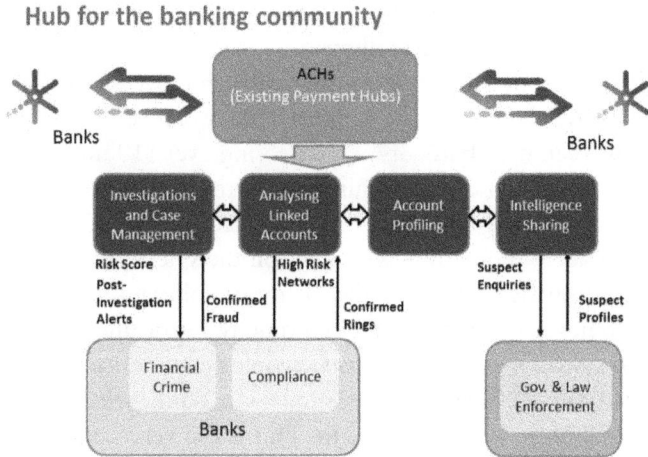

Hub for the banking community

The Hub enables a third party to look at both sides of the payment transaction and to which accounts they go to. Not only that, the transactions are then followed from account to account showing the network of transactions which more easily flags a fraudster or money launder.

The Hub also checks the compliance against the whole gambit of requirements. For example, OFAC compliance is shown. Today all the banks perform compliance checks to some degree. The fraudsters and criminals know who is best and when one of their scans works at one bank they go to another and see if that works.

Banks have always cooperated with each other especially when there are no competitive advantages. It makes sense for them to provide an industry utility to combat fraud and money laundering.

OFAC Analysis Team Process

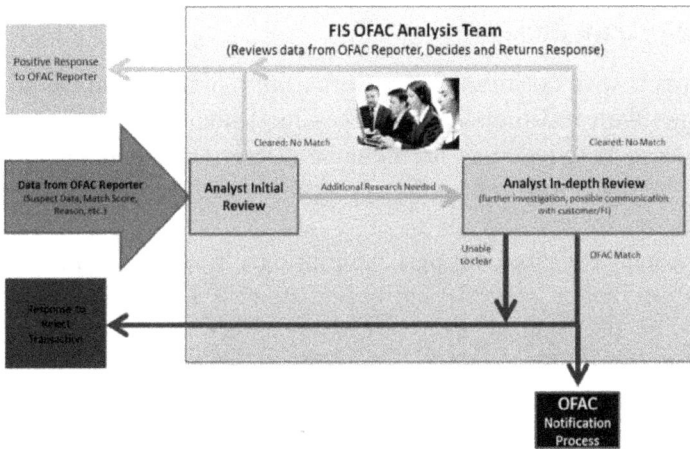

The big cry, especially amongst the fraudsters and money launderers, is that this is against the Data Protection Act. The Digital Policy Alliance, an all-party UK Parliamentary Group, engaged with all parties – the banks, the Government Departments, suppliers, fraud prevention organisations – found after extensive research it was perfectly fine for a 3rd party to handle the payment data from both sides. The 3rd party and the banks themselves have to follow procedures which are certified by the Information Commissioners Office. This includes making the data anonymous and the role of 'the data controller'.

It was found that the Data Protection Act permits the sharing of information under the right conditions. The DPA's act is to work further with the EU on the new Data Protection Act set for 2014.

The implications for fraudsters and launderers

Many banks and suppliers refer erroneously to the Act as reasons why certain frauds took place and they could not stop them.

There is also a movement underway by the authorities to make the banks use suppliers who are also bulletproof. Banks now have to make certain that their vendors meet the new standards of data protection and security.

Real-time alerts

THIS IS WHY COMPLEX EVENT MONITORING of business intelligence flows with real-time alerts is a key focal point. The ability for a bank to keep its finger on the pulse of every transaction across its global operations will be the key to protecting against internal and external threats.

And if real-time business monitoring can solve the first issue, an external or internal security breach, what do you do about the second area: ensuring ease-of-access securely?

Mobile banking apps from world leading banks like Wells Fargo, PayPal, Chase and others were failing basic security tests. At the time, August 2011, a quarter of all mobile bank apps failed basic security tests.[18]

Whether true of not, there are obvious flaws in mobile security right now, and yet there shouldn't be as mobile banking should be more secure than online banking. After all, the mobile phone is a useful authentication token, with the unique telephone number, ability to use text messaging for one time passwords, and even tracking the geolocation of the customer all being useful aspects of the system.

Customers will always know where their mobile is and have it with them, unlike their wallet or credit card, and are far more likely to know when it is lost or stolen. A famous story tells how a person

---

18   "Security Warning: 25% of Mobile Banking Apps Flunk Test", *American Banker*, August 2011

called the Emergency number and upon hearing that he had lost his mobile was told that it was not an emergency. He immediately started crying and said that it was a very real emergency for him because all his life was on it, including all his contacts.

The criminal's job is to continually test and try to break the security of the bank.

This means that the bank must therefore always be one step ahead of those who want to create cracks in their firewalls.

That means continual renewal of information security policies, systems and infrastructures, and making sure that the bank keeps up with the best practices in securing their customers' data. The key for the banks is should they handle this on their own or work with partners that can help prevent scams from working across the whole network.

Chapter 13: Banks as a digital police force

BANKS' LEGACY IT INFRASTRUCTURES are not designed for the collection of new information and transactions in digital form, as the software and computing is old. Core banking systems were only extensively revised to insure millennium compliance with the last great wave of core banking system purchases back in the 1990s.

The replacement of a core banking system is highly risky, and so the usual approach is to add supplemental systems. This is acceptable to most bank decision makers but, after a while, many bank systems look like a diagram of the solar system where the core is the Sun and the supplemental systems the planets. This means that everything revolves around an old legacy structure that is not fit for today's purposes.

For example, just the amount of information that is now required and continually asked for by the regulators is extensive particularly during client on-boarding, where basic due diligence information about the customer and the business relationship is sought.

It is at this stage that the bank needs a clear idea of who the individual is and what the company business is about. Often this is covered in the contacts between the bank and the company. Banks are obliged to verify the accuracy of the customer's explanation on the object and purpose of the company, so as to establish the credibility and accuracy of the information.

In addition due care is needed with a politically exposed person (PEP), who may be resident abroad. PEP is a person who has or had (normally in the past year) association with prominent public functions. These are public functions at the highest levels in a single country or at European or international level, such as Heads of State and government, ministers, etc.

Also, due diligence applies to people close to the PEP such as spouse, partner treated as spouse, parent or child (and his spouse) and known associates. Banks have found that the easiest and quickest way to obtain information about the customer's identity and ownership structure is to ask the customer for the required information and documentation. The documents routinely requested include:

- ❑ Certified copy of registration certificate not older than three months;
- ❑ Certified copy of share register – majority owners to undergo Know Your Customer process;
- ❑ Copy of the business plan;
- ❑ Copy of the budget;
- ❑ Number of persons employed;
- ❑ All authorised signatures need to prove their ID and address (passport and utility bill).

This is all wrapped in the Know Your Customer (KYC) process, to prove identity and that everything is above board. This information is required by the banks and monitored by the regulators. Hence digitisation can provide the information when the regulator requests it, especially when the banks need to provide these details quickly, as the movement of money is accelerating. This is best done through:

- ❑ **DOCUMENT MANAGEMENT** – The definition, scanning, management and tracking of customer documentation, and reporting of any deviations;

❑ ACCOUNT MONITORING – The tracking of movements within account(s) looking for deviations outside of a pre-determined profile;

❑ WATCH LIST CHECKING – Enter a name and the system will check to see if the name, or like sounding names, appear on any of the watch lists (e.g. OFAC, Bank of England and others); and

❑ MESSAGE MONITORING – Check all inbound and outbound messages, irrespective of format, to see whether any field (normally the Ordering Customer and Beneficiary) appears on one of the watch lists. Control over the granularity of the name checking so as not to create too many false alerts. Messages that fail Watch List Checking are put to a quarantine queue for manual intervention.

All the checks and actions taken have a full audit trail attached..

How digital banking allows the bank to avoid the underworld

As mentioned, the reason for all these KYC checks is down to regulations, many of which exist to stop money laundering. Money laundering fuels the drugs and terrorism business, and this is why governments want to ensure banks catch the criminals and when banks flout those rules, as Standard Chartered and HSBC were found to be doing by the US regulators recently, they get fined billions.

This is why Anti-Money Laundering (AML) and KYC checks are so important.

These requirements coupled with the Sarbanes-Oxley (SOX) and Dodd-Frank legislation in the USA means that to open a new bank account the client will need to have face-to-face meetings with the bank. In addition, payments going out or coming into an account are screened against 'watch lists'. Watch lists are names of companies and individuals wanted by various governments, and equally blocked countries where transactions are not allowed to take place. In this case, the USA has made it illegal for their accounts to be used for the funding of Iranian clients, and this is where Standard Chartered were proven lax as the bank, by law, has to continually review payment transactions.

Banks and credit companies are subject to criminal liability provisions of the law when they participate in certain financial transactions for one client. The new legislation provides for changes in a number of respects, including those relating to customer due diligence, providing information, compliance and enforcement. Money laundering legislation is important and difficult to apply. These regulations impose many obligations that must be observed by all concerned with or working with customers and money transactions in a company's business. It is important to work to prevent the company and its business being used for money laundering and terrorist financing and other serious crimes all the time is ????

The purpose of the policy and the instructions are to ensure that the company's anti-money laundering and terrorist financing and other serious crime is always adequate, appropriate and in accordance with current standards.

Under money-laundering legislation there are essentially three obligations to:

- ❏ Identify and verify their customers (due diligence);
- ❏ Review transactions that could reasonably be expected to constitute or be aimed at money laundering; and
- ❏ Report suspected cases of money laundering.

There is also the Financial Action Task Force (FATF) against money laundering, which was formed in 1989 by the G7 countries and continues its operations today. The organisation publishes many recommendations, forming the basis for much of the regulation concerning measures against money laundering, as well as updates to the lists of non-cooperative countries and territories (Non-Cooperative Countries and Territories, NCCT).

All of this demands systems change to keep up, especially in the area of payments, and changing the payments structures of the world is not easy for the reasons given earlier but for other, more fundamental reasons too.

Chapter 14: Digital Banks and the bank-to-client relationship

BANKS ALL HAVE FOLLOWINGS – look at the number of accounts as a guide – but unless they adapt to the digital age quickly, then that following may go elsewhere. It is what the music industry misunderstood and is just starting to deal with effectively.

These days, you don't sign music artists to write songs and sell records, you sign them to write songs and give them away for nothing so that people will follow their website, buy the T-shirts, come to the gigs and download the odd track.

This is why entertainment firms sign their artists up for 360-degree contracts these days – all the music sales and the rest – rather than for a recording contract.

This is what banks need to focus upon. It is all about providing the consumer and the corporate with some real value, such as aggregation services, lifestyle financing advice, real-time risk management, and helping them live a happier lifestyle by providing ongoing financial assistance.

This we now call PFM, Personal Financial Management.

PFM is a functionally rich set of financial tools that help consumers – and businesses – aggregate their financial transactions services across multiple providers into an advisory engine that can help them get better returns from their money.

PFM links you with other users who have financial behaviours like yours, and show you how to improve your financial returns based upon what people like you do. PFM might link to your mobile and social networks, allowing you to do a lot more intelligent financial structuring and operation. PFM can alert you to budgetary and balance issues, payments and billing notices, and interest saving or gaining opportunities.

Some want to focus on the social aspects of finance, whilst others on the financial management aspects; some want to provide offers and coupons, whilst others want to provide advice and analysis of your financial behaviours; some want to provide PFM online, on the mobile and on the tablet PC, whilst others only care about functionality rather than interface; and so on.

In other words, all of these PFM systems are slightly different with some easier to use than others, some more functionally rich than others, and some clearly in the lead over others.

It goes further than this however, as PFM combined with mobile provides real-time financial analytics and management for every individual and company being serviced by the bank.

Real-time and personal

THE SMARTPHONE IS WITH US 24/7, in our pockets and purses.

The screen never leaves us, and we can be contacted at any time of day or night.

With a mobile financial service we can have real-time PFM, showing our past – or passed – financial transactions, how they relate to our overall spend and more importantly what we can spend today, right now.

That's really powerful.

So I get my phone out to pay and the phone not only tells me the PFM piece about how many cappuccinos and paninis I've bought in Starbucks for the past month, week or year, but also whether I can afford it.

That may sound redundant to some of you, but if your phone can show you proactively your behaviours and habits financially and, in real-time, alerts about whether this next transaction will take you overdrawn, then that's really something.

Think about it in those old days when you were a student, hard-up or broke.

Every penny counts and every transaction can be a moment that sends you into the overdraft zone.

All those increasingly expensive fees and charges, often invisible to the client, can trigger the overdraft. This in turn creates another fee and before the client knows the fees are larger the amount being rejected.

But now you can have that cappuccino and panini, not only knowing that you spent £50 in Starbucks this month and have far too many paninis and cappuccinos, but also knowing you can afford it.

That is really individual and means that banks can send highly personalised attention automatically through alerts across a wide spread of products. Today you take out a loan for three years and the chances are, if you are a timely payer you will never hear from that bank during the whole 36 months. No congratulations, no here we'll extend your loan at a better rate as you are a good payer and no recognition that you are a good credit risk. Many banks offer better pricing for 'new customers only'.

The same is true for payments, investments, deposits and other products. One in 10 Direct Debits in the UK are for services and products no longer required. Does a bank review your direct debits for you? Well, digitally this could happen each year with the time of that DD being highlighted.

The corporate with a wide spread distribution, real-time portfolio and cash management positions for treasurers, could be created. The dashboard and suggestions by the bank would help improve allocation of resources; on-going analysis of market and credit movements; tracking of cash and comparisons to last month and an anticipation of revenues

The fact is that it is the value the bank adds that really locks in the customer and the corporate. The bank becomes important again and not isolated from the rhythm of personal and business life.

This will be part of the new digital banking model and replace those transactional banks that are just processors.

Banking on the cloud

CLOUD COMPUTING IS A WIDE AND DIVERSE OPERATION that has gained a panacea status of being all things to all people.

It's everything you want and, as a result, it's lost its meaning.

In its simplest form, if you run your bank on anyone's technology you might as well think of it as a cloud.

But the risk of losing scale, resilience, security and control is the core issue at a bank's heart, and they're not willing to take that risk with cloud, especially if no-one can define it.

Initially, mainframe computer power and then BPO and virtualisation created efficient computing capabilities for financial firms.

Now we are moving to an age where home-grown, propriety computing and applications matter much less.

Just like the iTunes App Store and Google's Gmail for consumers, who really don't care how it's done and who does it, as long it's there, banks will gradually move to clouds, be it their own or someone else's..

The art will then be as to how you put your apps and resilience together through the cloud, rather than how you build and manage your internal fortress.

By that time, computing and applications for banks will be just like Gmail and iTunes for consumers – just stuff you plug-and-play, and pick and choose from.

Like a smorgasbord of utilities, the trick will be to make your plate of edibles the most attractive to the target audience you are trying to reach and, by this time, we will all think of technology, software and infrastructure like electricity and the Internet – just something you plug into and don't care how it works.

This then leads on to the component-based bank.

The component-based bank

BANKS HAVE TRADITIONALLY been vertically integrated business- es, where the end-to-end process is offered to the customer as a complete package. Deposit accounts, trade finance, payments pro- cessing, loans, credit, mortgages, savings, investment and more are wrapped together as pieces of business.

Within each separate product stream, banks also integrate the end-to-end, so corporates run their payables and receivables, elec- tronic and paper, domestic and foreign exchange all through the one bank. The same is true for consumers who run their entire everything through the bank when it comes to money, payments and transactions.

This is the piece that is changing, as new entrants attach each part of banking. New entrants like Currency Cloud, which offersr foreign exchange processing, or eToro, which offers portfolio man- agement in investor services through social trading.

What is happening is that bank functionality is being broken apart, divided into its lowest common denominators and then re- constructed in new forms.

The best way to illustrate this is by looking at a specific piece of banking, such as payment processing.

Traditionally, this has been viewed as its own specific product silo, but it's not an individual product line. For most banks, it's an integral part of the bank operations. It's the glue that hooks the customer to the bank. It's the core of their offer.

This is why banks find it so hard to think of payments neutral- ly, objectively, dispassionately. Payments are not objective, they are emotional. For most banks, they are very emotional. Ask a banker to outsource core payments processing and they'll give you a look like you're the devil's spawn. It's just not done.

This is where it gets interesting as banking has historically been a vertically integrated industry, providing an end-to-end service that is wrapped around the customer and is incredibly difficult to unlock.

And yet it will be unlocked as banking is driven into a componentised industry where you have a payment app, a balance app, a cashflow app, a budgeting app, a fraud app and so on and so forth. We can see this already by the number of social money and payments organisations appearing such as iZettle, PayaTrader, mPowa, SumUp, PayLeven, Go Payment from Intuit, Square and more.

Each and every area of banking is being broken down, componentised and reconstructed into new forms and new business models by the new Digital Bank regime of cloud-based mobile-social data analytics.

Suffice it to say that the componentisation of banking and shift from vertical integration to horizontal components that can be put together as you like, is happening, and happening fast.

Banking becomes plug-and-play

AS THE DEVELOPMENTS MOVE FORWARD, banking simply becomes a smorgasbord of plug-and-play apps that the customer stitches together to suit their business and lifestyle needs.

For example, here's an illustration of banking with no bank involved.

Using prepaid cards, you can load and use a MasterCard and just keep topping up without ever using a bank. Equally, some cards provide prepaid without limits, with value being loaded by using Wire Transfers, Western Union, MoneyGram, E-Gold and other commodities to up $90,000 per month to be withdrawn through ATMs worldwide.

Alternatively, some people might use PayPal or other services for payments. Although PayPal needs a bank account to open your account – they run their on-boarding check through the banking system – once the account is up and running, the user could close their bank account and purely exist within PayPal via revolving credit.

What is happening under the digital age is that the old model of banking, where everything is packaged together around a deposit account with a cheque book, is bust.

This why some banks are starting to white label and break apart their traditional services, so that corporates can buy just the bits they need. That might be a SWIFT Gateway here, Internet payment service there, international money transfer here, cheque processing there ... all bits of banking, all priced and packaged as a plug-and-play service.

The component-based bank

THE COMPONENT-BASED BANK is a method of taking complex applications and offering them as web services where you pay for what you use. In other words, banks offer their capabilities for anyone to plug into their services from payments processing to balance checks.

The components themselves are widgets.

For banks, these widgets are offered as simple pick up and drop code for anyone to incorporate financial functionality into their service. A great example of a financial widget is the one that can be generated easily by PayPal users to take credit and debit card payments online. It takes about five minutes and then can be plugged into any website to receive money and, for small merchants, is a major bonus.

These widgets are plug-and-play for anyone wanting that piece of bank functionality or, if the firm that wants this functionality is more sophisticated, they might pick up the code as an Application Programming Interface (API).

Digital Banks and the bank-to-client relationship

HSBC was one of the early banks to adopt this approach with the then COO Ken Harvey stating in 2009 that he can launch the bank into any country simply and easily thanks to their technology structure. All the IT is free, he claimed, because HSBC build it once and then deploy globally through the network so that it can be used by thousands of staff and customers in almost 200 countries. That is the cloud component based bank in action, with one program built for thousands of users in hundreds of countries. But, and it's a big but, virtually every banking system had something different. Every system had to reflect the local regulatory requirements and local customerisation. The Midas system is in over 300 banking locations and each location is slightly different. These systems, especially in the global banks, cannot communicate with each other. Each is an Island.

HSBC, like Citi and many more, grew internationally under 'barons' who were held responsible for their profit and loss. The Barons acted in their own interest to make their numbers. They did not care about the IT architecture, in fact one global bank has virtually all the Core Banking systems that were available and also built their own. Bank of New York won an award for home grown desktop software which was better than Windows. Now in the digitally connected world how do you compete with an IT and processing cost structure that cannot be sustained.

Imagine a core banking system running at an international branch costing $10 million a year to run. One bank has 40 of these systems. That is $400 million a year just to stand still yet the the bulk of the revenue comes from a few countries.

The answer is how do you consolidate the existing IT for the digital bank. The Bank Board would then have to be involved and how many of the Board are in the digital age? I bet they all have smartphones which run the same regardless of geographic location.

The new goal for any bank is to combine enterprise business with technology. when redesigning their technology. Any bank should now be able to enter any country with a mortgage, credit card or deposit account product and the only costs are the people and physical investments in buildings. The rest is on the network. Unfortunately the legacy systems and Industrial age of mill owners (the Barons), at most banks, made it difficult for change to take place quickly.

And it's not limited to banking either, as the new European equities exchanges are demonstrating. Chi-X BATS Trading, NASDAQ and more are all launching radical new trading systems using leading-edge technologies at a tenth or more of the cost of the traditional exchanges.

They also have a tenth or less of the staff and overall, this is why these new trading venues can charge 10 basis points per side to clear and execute trades in under 2 milliseconds, compared to seven times the cost and many times the time taken by the incumbents.

Any area of financial services you want to point to, you can see new entrants and innovative incumbents changing the model by

1. The ability to grow with minimal additional costs – just adding more traffic to the network;
2. Componentise the bank into widgets that can be picked up and dropped by staff and customers as they feel;
3. The opportunity for staff and customers to create banking homepages that are completely personalised to them;
4. Flexible, comprehensive, highly tailored components.

Volume is the goal for bank building components because additional volume adds zero cost and purely feeds return on investment.

Market the widget heavily, build up the volumes fast and focus upon the service delivery – the human interfaces – as your critical value-add differentiation. Your value-add is how you package the widgets and present them, not the widgets themselves.

That's why Citigroup have been marketing their APIs, apps and other services heavily. They want volume on their widget, and Citi are one of the few banks who have been white labelling their systems to other banks.

Once you've built the infrastructure, it's all about getting volume because it costs little more to process a billion calls than it does to process a million.

A head of payments at one bank said: "Our technology guys asked me why we charge more for a $50 million payment than we do for a $5 payment, when the infrastructure costs to process are the same. Are they mad?"

No sir. They are asking an obvious question at the heart of the change that needs to be made to banking cultures, as they realise the change that technology now delivers.

This payments executive was from that old school of banking waking up to the new world realities.

Thirty years ago, when many senior bankers were starting out in their banks, they were told that technology was expensive, inflexible and must be used forever.

That's why every project was massive, time-consuming and demanded huge cost.

When SWIFT, MasterCard, Visa and the key networks for transaction processing were built, for example, they had to be built by an industry consortium. No individual bank could afford such a huge project or cost. That is why these were cooperative groups back in the 1970's across all banks.

Back in 1975, the bank technology economic model involved massive cost that could be recouped slowly over time through usage. Cost was depreciated slowly, and could only be covered by high prices and margins. Hence, banks, SWIFT and the card companies all worked hard to create the infrastructure, and cover the costs of that infrastructure through high transaction processing and interchange fees.

This model had changed a little by 2000.

What was happening was just after the 'Year 2000' fiasco was over. Many systems had been renovated and rationalised and the costs of building new systems had reduced somewhat thanks to HTML (an Internet language) and component-based modelling.

Therefore, banks found the time and cost of building new systems had been fundamentally reduced but still had a technology challenge, as the costs for usage and pricing did not come down as fast.

For example, the then CEO of SWIFT, Leonard Schrank, was interviewed in 2003, and made a few statements about SWIFTNet was well underway using IP-based messaging, costs for a typical SWIFT message had come down 70% in the last decade, whilst volume had increased four-fold to 8 million messages per day.

What we were seeing a decade ago was the cost to build was coming down, and volume was raising faster than before, thanks to ease of communication and connection.

Today, the cost to build has become virtually irrelevant if you are using the right tools but, once built, the volume can be increased really fast, thanks to openness, standards, ease of networking and communicating.

In other words, the economics of banking has fundamentally changed. This is because banking is based upon technology, and the economics of technology has fundamentally changed.

The new economics of banking: technology and systems, processing and functionality are virtually free; anything that is commodity activity should be brought in as a service; and all pricing is for value-add, not commodities.

Bearing in mind that banks are now offering components and customers will integrate and incorporate those components as they see fit, another change will occur. Banks will collaboratively compete. Sometimes they will be partners and, at other times, they will be competitors and this will not just be between banks but between banks and telecommunications firms, technology vendors, information providers and more.

When Lehmans collapsed in September 2008, it was the week of the one of the biggest banking conferences in the world, SIBOS in Vienna. SIBOS had various themes and finished on the Thursday with a presentation by Don Tapscott, the author of the book *Wikinomics.*

Don then made this appeal at the end of his presentation:

"The risks in the financial system must be better managed in the future

so why don't we create an open source group for risk managers? A Facebook style system the, with likes and pokes for Risk Professionals community to exchange experiences. This group could then share and discuss risks in the financial systems and have contributions from all. Effectively, risk management becomes an open source arena so that everyone can build a more robust, reliable and resilient future."

It seems we're all waiting for the regulators to come up with their plans before implementing ours, but isn't that wrong?

Collaborative competition says that for things that are commodity, or things that are industry wide issues and infrastructures, we should just widgetise them and make them plug-and-play. This processing capability and knowledge should then be made available to all.

Then, as a bank, you focus upon the areas where you differentiate. These are the customer-centric parts of engagement, acquisition, delivery and fulfilment.

The result is that we could outsource compliance and create common shared components for the Anti-Money Laundering (AML) and Know Your Client (KYC) regulatory requirements for account opening. AML and KYC just become a widget of functionality that you plug-and-play.

We could drop a widget into our system for payments processing too – just use a white labelled processor –- and even find a widget for a bit of credit risk management based upon an open source structure.

The rest – all the processing, compliance, risk and ancillary products – we've just dropped in as widgets of functionality into our banking structure. And those bits are all the bits of collaborative competition therefore.

We may compete with the providers of our widgets, but we also collaborate with them to use their processing where it makes sense or where it makes our own infrastructures more robust and reliable.

A quarter of a century ago, car manufacturers prided themselves on having the best manufacturing.

They produced all of the car's components and the manufacturer with the best components offered the most expensive cars.

Today, nearly all cars are based upon standardised and commoditised manufacturing of the pieces.

The manufacturer no longer manufactures, but just assembles the pieces into the whole and adds their own unique recipe of chassis and engine to differentiate.

All the components they just get off the assembly line of banking functionality, but the banks that assemble them to address a specific target audience in the most appropriate way will win that audience's business.

Chapter 15: Strategies to become a Digital Bank

IF WE WERE CREATING A BANK TODAY, how would we create it?

Building a bank today requires many things and so the first thing we would need to start with is what sort of feeling we want the bank to convey.

Are we a technobank or a human bank?

Do we want to encourage human interactivity or remote interactivity?

How do we believe we are different and what can we deliver to clients to show that we are different?

Where the weak points in the current bank offer, and how do we exploit those?

So our first point of focus would be how to create a cool bank.

The Apple of banking

TO CREATE A NEW BANK that was cool like Apple, we would start with an amazing online user experience designed for mobile and tablet computers.

A cool app store of financial offerings.

Designing a Digital Bank based upon Internet Protocol should be wholly focused upon today's mobile Internet. Being a bank it needs gravitas. People and businesses want help on financial matters. In the UK 20% of the population would like help in budgeting and running budgets of their financials. Therefore like models for cosmetics, the digital banks need an image that conveys the bank's integrity and trust.

Offer apps that are PFM – Personal and Corporate Financial Management – capable and making them simple to use online and on mobile.

Making security cooler by authenticating by using geolocation and signal ????

Making it simply cool by never asking me for names, account numbers and passwords, but simply giving me a personal space where my voice activates the services.

Making it truly cool by offering gifts and goods for loyalty, such as flights and iTunes based upon account usage.

Making it way cool by offering me those gifts and goods as I Google to buy them or walk past the branch instead of asking the teller if I needed home insurance.

And making it cool by relating to me based upon how I want to relate to you. This means talking to me as a human when I call you, and making me feel you really understand me because you've understood my data.

Giving it gravitas by using people that are fiscally believable.

Someone they could trust.

A cool Apple style bank offer that is mobile app-centric and celebrity endorsed by a trustworthy figure – next how to organise and architect the bank and our cultural approach.

Creating a cool and fair bank

I HAVE NO TARGET CLIENT BASE EXCEPT ONE: those who want to deal with a cool and fair bank through the mobile Internet.

My bank's customer is one that wants to deal with a 21st century bank in a cool and fair way, and to be treated like a human.

My marketing programme would therefore focus upon using social interaction online to attract viral amplification. It would make it clear what cool and fair banking over the mobile Internet means, and how to engage with us. It would Twitter, LinkedIn, Facebook and YouTube its services to build a fan base. It would dialogue online in a hugely human way. And it would be transparent in fees and approach.

It would build its processes based upon the customer outside-in view of the interactions and user experience people desire, and it would target to overcome the things that upset most people, such as lock-in fees, hidden charges, balloon payments on overdrafts and charges for the mundane.

The culture would be defined this way, with the aim of creating a happy culture of cool and fair people who get the mobile Internet. This happy culture would create happy customers. Customers who like cool and fair banking from a bank that gets the mobile Internet.

A cool bank is a bank that people want to be with, either as a customer or working for the bank.

Why incumbent banks could disappear

ONE OF THE BIGGEST THINGS that constrain existing banks is their past. Most were created over a century ago to manage retail and corporate transactions in branches, over the counter with human tellers and, to a large extent, this is pretty much how most branch-based banks look and behave today. Compounded by those banks that went international at the bequest of their corporate client who created what they had at home in the new country.

Yes, these banks have added new functionality over the years, and new channels, but these banks have not overcome this branch-centric view in that time.

Now those banks are changing. They are trying to become Digital Banks.

This is incredibly hard because the banks are huge monoliths, with hundreds of thousands of staff and heavy investments in existing operations and infrastructure. Turning this around to a nimble and fast Digital Bank is proving nigh impossible.

If you take the UK's banks, none have replaced their core systems in recent times. Instead, they have tried to evolve them and a replacement has only ever occurred when it has been forced through merger or acquisition, such as those of HBOS and Lloyds or RBS and NatWest. In both instances, the banks migrated customers from one banks' legacy system to another legacy system.

Wholesale replacement of core systems is difficult. This is amply demonstrated by the migration of Abbey and Alliance & Leicester to Santander's core systems over the past few years, and the Co-operative writing off $500 million on a new core system in 2013 without it being implemented.

In fact, it has been noteworthy that many banks and exchanges have been facing computer failures recently. The best known one was the RBS glitch of 2012 where their core payments systems were out for almost a month, but there have been many more since 2010. The key is old technology that cannot cope with the modern world.

Banks have many legacy systems across their core back office operations and so it is far easier to change and add new front office systems – new trading desks, new channels or new customer service operations – than to replace core back office platforms – deposit account processing, post-trade services and payment systems

Because the core processing needs to be highly resilient; 99.9999999999999999999999% and a few more 9's fault-tolerant; and running 24/7.

In other words these systems are non-stop and would highly expose the bank to failure if they stop working.

It is these systems that cause most of the challenges for a bank, however.

This is because, being a core system, they were often developed in the 1960s and 1970s.

Back then, computing technologies were based upon lines of code fed into the machine through packs and packs of punched cards.

The cards would take years to program and days to update in batch.

Tens of thousands of lines of code would inter-relate in modules that would mean any change to any minutiae in any single line of code would rip through the rest of the program and potentially corrupt it.

That's why banks would not change or touch these systems and is the reason why, once they were up and running and working, they would be left to run and work non-stop.

Then the world moved on, and technology became a rapidfire world of consumer focused technologies. Add to this the regulatory regime change, which would force banks to respond more and more rapidly to new requirements, and the old technologies could not keep up.

Finally, the technology has to change.

A competent bank de-risks the risk of change by testing, testing and testing.

But the real issue with an upgrade or consolidation is that it has be done more and more frequently due to the combined forces of regulatory, technology and customer change.

The mobile Internet world squeezes and exposes the legacy on the one hand – this is why many banks have struggled to incorporate mobile services with their Internet banking services – while the global, European and national regulatory requirements are placing further pressures on the core processes as well.

Therefore banks avoid migration to new core systems, and are handcuffed into legacy operations through their legacy systems.

Apple always has been cool, yet Microsoft is nerdy.

Because Apple is designed for people to use while Microsoft is functional and requires the owner to read the manual.

So to be a cool retail bank, you have to be different, creative, aspirational and accessible, and very consumer focused and, in this century of mobile broadband, you have to be as switched on as your customers.

Conclusions: the Digital Bank of 2020

IMAGINE IT'S 3RD JUNE 1838.

You are a canal owner and have just completed digging the waterway from London to Bristol, a distance of 200 miles. It has taken years and, in today's dollars, cost billions.

On 4th June 1838, Isambard Kingdom Brunel opens the Great Western Railway to link the same two cities by rail. The canal boats travel at five miles per hour, whilst Brunel's trains speed at 35 miles an hour. Shipping goods to and from the cities took four days by boat compared with six hours by train.

How would you feel?

Now imagine you are a bank and have been in business for over a century. The movement of money has taken days and still does. The usual answer to anything that requires fast decision-making is no, as banks need to avoid risk. Various fees are taken at source, often at the surprise of the customer, and the transition from paper processing to electronic is taking an age. Then technology allows real-time money movement that is fully transparent, with decision making in minutes.

How do you feel?

The banking owners are facing a similar dilemma to the canal owners. Not surprising the train immediately took all the business between the two cities. Similarly today, the digital age has arrived with the smartphone, and banking and their suppliers need to move out of the pre-Internet era.

Brunel heralded the Industrial revolution and changed the world forever. Now the Digital Revolution is doing the same thing, but on a global scale. The battleground is how we approach the future empowered with technology available everywhere. The smartphone is only going to grow in computer capacity and capabilities. People everywhere, not just the Industrial Age developed counties, will be able to buy and sell to anyone regardless of geography.

According to a 2012 study by Yale University[19], the average lifespan of a company listed in the S&P 500 index of leading US companies has decreased by more than 50 years in the last century, from 67 years in the 1920s to just 15 years today; and yet, in most countries, the banks have been established for over a century and rarely does a new competitor come into those markets.

For example, when Metro Bank opened in the UK in 2010, it was widely remarked that this was the first new retail bank in Britain for over a century.

When you review the largest banks in the world over a decade there is a consistency of names that now reflects the importance of Asia. For example, according to _The Banker_ magazine, the top five banks in the world in 1999 were:
1.   Citigroup
2.   Bank of America
3.   HSBC
4.   Crédit Agricole
5.   Chase Manhattan
In 2013, the top 5 are:
1.   Industrial and Commercial Bank of China
2.   JP Morgan Chase
3.   Bank of America
4.   HSBC

---

19   _Can a company live forever?_ BBC, January 2012

5.   China Construction Bank

This is because banks are integral to commerce and the economies of countries. The bigger the economy, the bigger the bank and the US and China are amongst the world's biggest. This has been demonstrated time and time again, no better than in the recent crisis. Banks can cripple economies or enable growth and progress, and their role is obviously to support the latter than create the former.

You need to have a licence to operate the fundamental of banking: deposit holding.

That does not mean that new banks and new forms of commerce will not build on the banking system, but it does mean that the core of banking has to remain with banks. Banks that are licensed, comply with state laws, are subject to auditing and supervisory interference.

These unique capabilities of the industry – an industry that in many ways is almost a nationalised service (but just not run in a nationalised way as that would give governments too much incline to corrupt) – make it one that is the foundation of all commerce, including new forms of commerce; but things around that core are changing, and changing fast.

Taking into account that the world needs banks to be licensed, as a fundamental, provides some view towards the future based upon the PayPal model.

PayPal began as a pure person-to-person viral payment mechanism and, over time, has matured into an established player in banking. The firm has a banking licence, but prefers to work with banks to build their business as a pure payments processing focused organisation.

This does not mean that PayPal will not change this premise and become a bank one day but, right now, they are the only new non-bank play of any note to have made inroads into the financial system and they have achieved it all on top of the system.

PayPal's play being to make banking and payment easier by

wrapping the process into an easy way to get the thing you want, rather than having to think about paying for the thing you want.

Amazon has done the same with clicktopay as has Apple with their aggregation of microtransactions through iTunes and apps. However, in the case of Apple they could change the world later this year when their new operating system for the iPhone and iPad incorporates a mobile wallet. The Apple Passport is a virtual wallet that, once activated, will provide integrated coupons, loyalty programs and other offers alongside proximity and NFC payments. Google had already introduced such a wallet in a partnership with MasterCard and Citibank last year, but it has not taken off yet.

The reason being attributed to a lack of critical mass due to Google's decision to limit their virtual wallet to only one type of handset. Conversely, Apple with their smartphone dominance globally will see 100 million handsets automatically activated with virtual wallets linked to 400 million iTunes accounts. That is critical mass.

Again, like PayPal and Google however, the model is one that leverages the bank's system rather than replaces it.

This is also true of other new entrants, such as Movenbank about to launch in the US. Movenbank aims to create a truly social mobile financial service, where gamification, personal financial management, social media, social networking, payments processing and traditional banking all come together in a simply integrated form.

Although much of the financial ecosystem remains within the bank realm, there are some signs of change in this new world, particularly where people are connected 1:1 via the mobile Internet. Especially in emerging markets where they have no legacy systems to protect.

M-PESA is an example of a near-bank system that transacts 20% of Kenya's GDP outside the banking system. Since M-PESA launched in 2007, the number of people who now have bank accounts with the traditional banks has almost quadrupled.

The reason?

As mobile money provides inclusion for the unbanked and under banked, they become recognised as financially viable citizens and therefore for financial inclusion by being banked.

Similarly, when banks find they cannot offer services to customers through this credit crisis that were previously offered, such as credit, other services gain traction. This is clear from the rise of crowd funding platforms like Kickstarter and social lending services like Zopa, Smava and Prosper. The crisis also brought to the front Payday Loans, such as Wonga, where interest rates charged would make loan sharks blush.

Similarly, savers who were receiving poor interest rates sought alternative investment avenues, and also found Zopa where interest on savings is typically higher than with a traditional bank. Zopa now controlling two percept of the UK personal credit market today, managing around £200 million of funds. Lending and saving are a permanent theme throughout the world. The status of where we are between poverty and wealth is in constant flux. Life events change the position constantly and tales of rags to riches and riches to rags are common. What has changed is the access by anyone to either lending or saving at any time. The issue for most is the pay day loans and draconian charges by banks for unauthorised credit. This type of unregulated charges led to 'The Debt Trap'. Once here for either individuals or companies it is extremely difficult to leave.

Poverty to Wealth

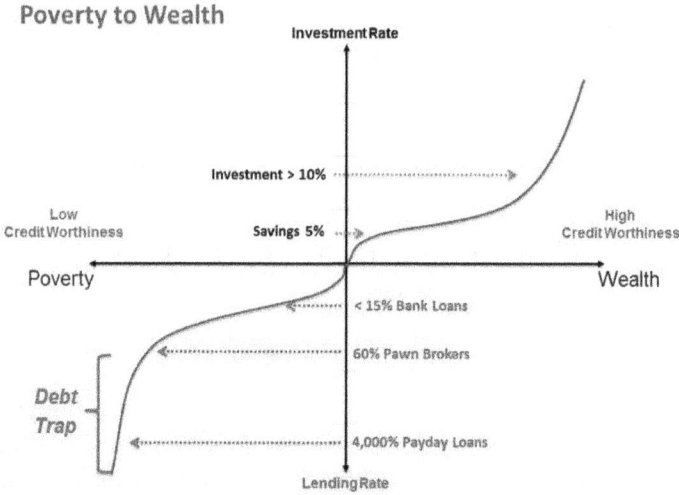

M-PESA, Zopa and Wonga are purely more mature versions of existing services that banks could offer and then we see more change, thanks to virtual currencies such as Bitcoin.

The idea of a Bitcoin is to enable anyone to use a digitally encrypted exchange of value, which can be purchased through recognised exchanges using traditional currencies. Once purchased, they can be traded worldwide without any cross-border fund transfer charges.

This has not taken off as yet, but if Bitcoins were to be traded like QQ currency, then we would have a dramatic change to the whole basis of finance.

QQ is a form of digital currency that has been traded in China for over a decade. Originally created to allow users of the Internet to buy ringtones and download games, it has grown into an accepted form of digital exchange for any online goods and services.

QQ is now recognised by the Chinese government – providing information to the police about crimes is now rewarded with Yuan or QQ coins – and has become a real currency.

Many other examples of such currencies are appearing daily – World of Warcraft or Diablo Gold, Flips, Zynga credits and more – and it is only a matter of time before banks will be managing and trading virtual currencies alongside traditional money.

New forms of money – Bitcoins or Zynga credits – are managed by new forms of bank (often without a banking licence) then potentially a new form of player will purely manage the new form of commerce.

That is not a disaster, but it does change the financial ecosystem, as the virtual wallets will be hybrid with the real wallet, just as the virtual world is becoming hybrid with the real world.

This hybrid world in which we live is enhanced by a digitised augmented reality. This means that we really need to stop thinking of the world as channels – branches, call centres, Internet and mobile – and just think of it as a digitally enhanced, augmented financial experience.

That is what the new players have fundamentally realised and if banks and traditional players ignore this fundamental, then they will not only lose out on the new forms of commerce, but on the old ones too.

At the same time the banking infrastructure, built over many years, is geared to move money between the banks at a pre-Internet pace. This is especially true in the developed economies with the emerging countries able to harness digital banking immediately. The magic will be to upgrade the infrastructure to meet the demands of the digital age, without losing the business.

The digital age is here, so let's all embrace it.